NONDESTRUCTIVE TESTING METHODS FOR CIVIL INFRASTRUCTURE

A collection of expanded papers on nondestructive testing from Structures Congress '93

Approved for publication by the Structural Division of the American Society of Civil Engineering

Edited by Hota V.S. GangaRao

Published by the
American Society of Civil Engineers
345 East 47th Street
New York, New York 10017-2398

ABSTRACT

This proceedings, Nondestructive Testing Methods for Civil Infrastructure, contains papers presented in the sessions on nondestructive testing (NDT) for the 1993 Structures Congress held in Irvine, California on April 19-21, 1993. The purpose of this proceedings is to bring the modern NDT techniques that are being used in the aerospace and medical industries into the civil infrastructure. To this purpose, these papers deal with new developments of NDT methods and experiences for testing of materials, building components, and highway structures. Some specific topics covered are vibration monitoring, acoustic emissions, and ultrasonics.

Library of Congress Cataloging-in-Publication Data

Nondestructive testing methods for civil infrastructure : a collection of
    expanded papers on nondestructive testing from Structures
    Congress '93 : approved for the publication by the Structural
    Division of the American Society of Civil Engineers / edited by Hota
    V.S. GangaRao.
        p.   cm.
    Includes indexes.
    ISBN 0-7844-0131-4
        1. Non-destructive testing. I. GangaRao, Hota V. S.II.
    Structures Congress '93 (1993 : Irvine, Calif.) III. American Society
    of Civil Engineers. Structural Division.
    TA417.2.N677   1995   95-36308
    624'.028'7—dc20   CIP

FOREWORD

The papers included in the following proceedings are the full-length papers presented in the sessions on nondestructive testing (NDT) for the structures congress 1993 held in Irvine, California on April 19–21, 1993. Each of the papers included in these proceedings has received two positive peer reviews. All these papers are eligible for publication in the ASCE Journal of Structural Engineering.

While it is apparent that the aerospace industry has received more attention than the civil infrastructure in the application of NDT, the civil infrastructure including highway bridges and pavements require new technology or improvement of existing technology in terms of longer service-life to provide reliable quantitative information to insure the safety of our structures. Because of the neglect, infrastructure deterioration rates have led to productivity losses, user inconveniences, and severe decrease in ratings or load limitations. Hopefully, the use of modern NDT techniques can alleviate some of these problems. The purpose of these proceedings is to bring in the modern NDT techniques that are being used in the aerospace and medical industries into the civil infrastructure. To meet the above purpose, this document includes technical papers dealing with new developments of NDT methods and experiences for testing of materials, building components, and highway structures.

The focus of these proceedings is to increase the awareness of the various nondestructive evaluation methods that are now the subject of research of material science and engineering.

The research issues addressed herein are strength, deformability, chemical degradation, and fracture of structural materials, components, and systems. The goals are to predict, control, and improve the integrity of materials in service and prevent catastrophic failures.

The research challenges do occur commonly in sensor technology for making the necessary measurements (*nano and micro level*), sometimes under hostile field conditions and with limited access. Also, NDT research demands on quantification of nondestructive evaluation signals so that the information about the state of the material provided by such techniques can be used with confidence in condition assessment and remaining life estimates of a facility. The topics discussed in these proceedings include vibration monitoring, acoustic emissions, ultrasonics, and others.

Hota V. S. GangaRao, Director, Professor,
West Virginia University,
Morgantown, West Virginia

TABLE OF CONTENTS

Contributed Papers

MODAL ANALYSIS TECHNIQUE FOR BRIDGE DAMAGE DETECTION

K.C. Chang[1], A.M., Z. Shen[2], S.M., and G.C. Lee[3], M., ASCE

Abstract

The dynamic responses of a wide-flange steel beam with artificially introduced cracks were studied analytically and experimentally. Frequencies, displacement mode shapes (DMS), and strain mode shapes (SMS) are determined in both the analytical and experimental analyses. Modal damping ratios are also extracted in the experimental study. The sensitivities of the change of the modal parameters due to the damages are studied. The absolute changes in mode shapes were used to determine damage locations. Results show that the damage of a beam can be detected and located by studying the changes in its dynamic characteristics. SMS shows higher sensitivity to local damage than DMS does.

Introduction

The modal parameters of a structure are functions of its physical properties (mass, stiffness, and damping). Structural damage will result in changes of the dynamic properties [Mazurek and DeWolf 1990, M. Biswas et al. 1989, Salane and Baldwin Jr. 1990, and Yao et al. 1992]. Therefore, damages to the structure in general will result in changes of the physical properties of the structure, and hence the modal parameters. Presently, measuring and analyzing dynamic response data have been recognized as a potential method for determining structural deterioration.

---

[1] Professor, Department of Civil Engineering, National Taiwan University, Taipei, Taiwan. (Formally of Department of Civil Engineering, State University of New York at Buffalo)

[2] Graduate Research Assistant, Department of Civil Engineering, State University of New York at Buffalo, Buffalo, NY 14260

[3] Professor and Dean, School of Engineering and Applied Science, State University of New York at Buffalo, Buffalo, NY 14260

1

Fatigue cracks constitute the most common reason for stiffness degradation of steel bridges. However, the changes in frequencies, damping ratios, and DMS associated with the development of these cracks are minimal and are difficult to distinguish from experimental noise. In this paper, SMS was used for damage detection of girder bridges. The rational for using SMS for structural diagnosis is as follows: Structural damage will always result in stress and strain redistribution. The percent of the changes in the stresses and strains will be highest in the vicinity of the damage, and hence the damage zone can be identified. An experimental study was conducted by using a model girder bridge. The changes in DMS, SMS, natural frequencies, and modal damping were recorded simultaneously as various cracks were introduced to the girder. A finite element model was also developed to obtain analytical results so that a comparison could be made with the experimentally observed data.

Theoretical Bases of Modal Analysis

The basic concept of analytical and experimental modal analyses was developed by Bishop and Gladwell [1963], Clough and Penzien [1975], Ewins [1986] and Bemasconi and Ewins [1989].

For an N-Degree-Of-Freedom system, the general equation of motion may be written as:

$$[m]\{\ddot{x}(t)\}+[c]\{\dot{x}(t)\}+[k]\{x(t)\}=\{f(t)\} \quad (1)$$

where [m], [c], and [k] are the N x N, mass, viscous damping, and stiffness matrices, respectively. $\{x(t)\}$ and $\{f(t)\}$ are the N x 1 vectors of time-varying displacements and forces.

Suppose a proportionally damped structure is excited at point p with the responses recorded at point q, the component of the Frequency Response Function (FRF), h_{qp} is given by:

$$h_{qp}=\sum_{r=1}^{N} [_r Y] \phi_{qr}\phi_{pr}=\sum_{r=1}^{N} \frac{\phi_{qr}\phi_{pr}}{K_r-\omega^2 M_r+i\omega C_r} \quad (2)$$

where Φ is the component of the mode-shape matrix $[\Phi]$ and

$$[_r Y] = \left\{\frac{1}{K_r-\omega^2 M_r+i\omega C_r}\right\}_{diag} \quad (3)$$

is an N x N diagonal matrix. In Eq. 3, M_r, C_r, and K_r are the components of the generalized matrices [M], [C], and [K] respectively.

The strain field may be defined as follows:

$$[\psi] = [D] [\phi] \quad (4)$$

where $[\psi]$ is the matrix of strain mode shapes, [D] is an N x N matrix of linear differential operator which translates the displacement field to the strain field, and

[Φ] is the DMS matrix.

The general expression for the components of the Strain Frequency Response Function (SFRF) and then be expressed as:

$$S_{\epsilon p}(\omega) = \sum_{r=1}^{N} \frac{\psi_{qr}\phi_{pr}}{K_r - \omega^2 M_r + i\omega C_r} \quad (5)$$

where

$$\psi_{qr} = \sum_{j=1}^{N} D_{qj}\phi_{jr} \quad (6)$$

It is clear from Eq. 2 and Eq. 5 that in experimental crack simulation, the displacement and strain mode shapes corresponding to different modes can be determined from the resonant magnitudes of different points on the Frequency Response Function curves.

After obtaining the FRF, the real and imaginary parts are extracted. Circle-fit analysis is then used to obtain the modal parameters. A set of measured data points around the resonance at ω_r is used for the circle fit. The modal parameters can be obtained from the modal circles.

Referring to Fig. 1, the damping of the mode can be obtained by:

$$\xi_r = (\omega_a^2 - \omega_b^2) / (2\omega_r^2(\tan(\theta_a/2) + \tan(\theta_b/2))) \quad (7)$$

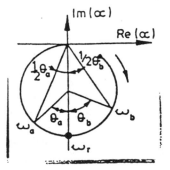

Fig.1 Fitting Circle

Where, ω_b is a frequency below the natural frequency, ω_a is a frequency above the natural frequency, and θ_b and θ_a are related phase angles.

The natural frequencies are the values which maximize the following expression:

$$\frac{d\omega^2}{d\theta} = (-\omega_r^2 \xi_r)(1+(1-(\omega/\omega_r)^2)^2/2\xi_r) \quad (8)$$

In Eq. 8, θ is the phase angle, Z is the damping ratio of the rth mode.

The mode shapes can be obtained by observing the diameters of the fitted circles at all measuring stations. They are then normalized with respect to a reference station [Liang and Lee 1991].

Experimental Setup and Test Procedure

A standard W6X20 steel I-beam with a 12-foot length was used as a model girder bridge in this experimental and analytical study. Fig. 2 is a schematic drawing of the test specimen. The end supports were two hinges connected to the bottom flange of the beam. These supports restrained only the longitudinal and vertical motions. The direction of the introduced vibration was in the plane of the web.

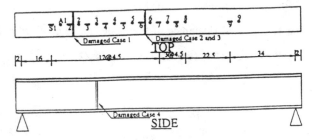

Fig. 2 Layout of Specimen, Cracks and Measuring Stations in Test

Four different damage types were introduced to the beam. Case 1 was a full flange cut located between A1 and A2 at 2.3 inches from A2. Cases 2 and 3 were half deep flange cut and full deep flange cut located between A5 and A6 at 2.3 inches from A6. Case 4 was a vertical cut on the web with a depth of 6 inches (full web height) at A3. The width of the cracks was 1/16 inches introduced by an electric saw in the specimen.

The locations of the accelerometer and strain gage stations are also show. Accelerometers are identified by A1, A2, etc. and strain gages by S1, S2, etc. Since damages were designed to occur between stations A1 to A6, A9 is selected as the reference station for all accelerometer stations and S9 for strain gage stations in the normalization of the mode shapes.

A 12-pound impact hammer was used to excite the test structure. The data sampling rate was 600 Hz.

Tests sequenced from Cases 1 to 4. At the beginning of a test, the baseline signature was measured on the undamaged beam, then a crack was cut and the dynamic characteristics associated with the damaged beam were determined. After the damage Case 1 and 3 (the full flange cut) were completed, the cracks were welded, and the signature from the "repaired" beam was redefined as a new baseline for the next test case.

In every test case, force and responses of 20 strikes were recorded for analysis. The digitized signals were Fourier transformed. An averaged-frequency response function (FRF) was calculated from averaged power and cross-spectrum for each channel. On every Fourier transform, a total of 4096 points were used and the resulting frequency resolution was 0.1465 Hz. After getting the FRF, the modal parameters were obtained by the circle-fit method [Ewins 1986].

To examine the accuracy of the test, the coherence function and the statistical analysis of frequencies and damping ratios were considered based on the data extracted from all sample stations. The mean value, standard deviation (σ), and coefficient of variation COV (σ/mean) were calculated.

Owing to the limit of the impact hammer, only the first mode of the beam is clearly excited. The following discussion pertains to the variations of the first mode response.

Analytical Crack Simulation

Modal analysis of the finite element model of the test specimen was considered to compare it the experimental observations. This analysis was conducted by using "ANSYS".

Finite element models using solid elements (Fig. 3) were generated for both the intact and the damaged beams. The undamped natural frequencies, displacement mode shapes and strain mode shapes associated with the first vibration mode in the plane of the web were calculated. In order to compare mode shapes for different damaged cases with the mode shape of the intact beam, a reference station is necessary. Since the damage was designed to occur on the left-hand-side of the beam, station 25 is selected as the reference station for all the cases. Eigen value analyses were performed for the intact and damaged models were performed to obtain the natural frequencies and DMS in the vertical direction. SMS can be analytically predicted by imposing values of DMS on the model through static analysis. The resulting strain values would be the SMS.

The finite element models for the intact structure and the three damaged cases are given in Fig. 3. Stations along the beam from which the data were abstracted are also shown. Damaged Cases 1 and 3 were on the top flange with the cracks located between station 7 and 8 at 2.2 inches from station 7 for Case 1, and between station 15 and 16 at 2.2 inches from station 15 for Case 3. Damaged Case 2 was not simulated by the finite element model. Solid elements with 1/16 inch width were generated at the damaged locations and the relative elements were removed to simulate the cracks. The nodes located 2 inches away from the ends on the bottom flange were modeled as simply supported to imitate the laboratory model.

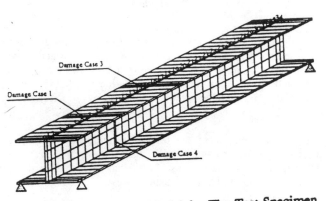

Fig. 3 Finite Element Model for The Test Specimen

Randomness of Dynamic Response and Test Accuracy

Because the experimental data contain certain noises and other experimental error, the measured responses possessed a certain degree of inaccuracy. To determine the experimental accuracy, statistical analysis is performed. In the circle-fit analysis, the modal frequencies and damping ratios are extracted from every sampling station in each test case. Tables 1 and 2 contain observed modal natural frequencies, modal damping ratios from accelerometers along with their mean, standard deviation (σ) and coefficient of variation COV.

As can be seen from Table 1, the maximum difference of the measured frequency was 0.1465 Hz, which is the amount of frequency resolution with the maximum standard deviation of 0.0772 Hz and maximum coefficient of variation of 0.0016. These results show that the error range in the measured frequency is approximately ± 0.0732 Hz (in one frequency resolution).

In Table 2, measured modal damping ratios are given. The maximum variations occurred in Damage Case 2 (the maximum change was as high as 0.00131) which has a coefficient of variation as high as 0.02403. Comparing the COVs with the data in modal frequency, modal damping ratios have a much higher variation than that of the modal frequency.

Table 1. Random Variation of Measured Modal Frequency(in Hz) from Accelerometers

STA.	CASES						
	C1 BL	CASE 1	C2,3 BL	CASE 2	CASE 3	C4 BL	CASE 4
A1	49.8047	48.3398	48.7793	48.9258	45.2637	51.1231	51.4160
A2	49.8047	48.3398	48.7793	48.9258	45.2637	51.1231	51.4160
A3	49.8047	48.3398	48.7793	48.7793	45.1172	51.1231	51.4160
A4	49.8047	48.3398	48.6328	48.9258	45.1172	51.1231	51.4160
A5	49.8047	48.3398	48.6328	48.7793	45.1172	51.1231	51.4160
A6	49.8047	48.3398	48.6328	48.7793	45.1172	51.1231	51.4160
A7	49.8047	48.1934	48.6328	48.7793	45.1172	51.1231	51.4160
A8	49.8047	48.3398	48.6328	48.7793	45.1172	51.1231	51.4160
A9	49.8047	48.3398	48.7793	48.7793	45.1172	51.1231	51.4160
MEAN	49.8047	48.3235	48.6979	48.8281	45.1498	51.1231	51.4160
σ	0.0000	0.0488	0.0772	0.0733	0.0646	0.0000	0.0000
COV	0.0000	0.0010	0.0016	0.0015	0.0014	0.0000	0.0000

Table 2. Random Variation of Measured Modal Damping Ratios from Accelerometers

STA.	CASES						
	C1 BL	CASE 1	C2,3 BL	CASE 2	CASE 3	C4 BL	CASE 4
A1	0.01811	0.01779	0.02199	0.02058	0.02239	0.01337	0.01210
A2	0.01795	0.01794	0.02144	0.02131	0.02224	0.01326	0.01199
A3	0.01875	0.01845	0.02211	0.02189	0.02248	0.01339	0.01218
A4	0.01927	0.01860	0.02203	0.02092	0.02326	0.01343	0.01220
A5	0.01919	0.01858	0.02195	0.02040	0.02353	0.01342	0.01218
A6	0.01919	0.01856	0.02175	0.02047	0.02333	0.01341	0.01216
A7	0.01925	0.01864	0.02172	0.02046	0.02335	0.01340	0.01220
A8	0.01921	0.01857	0.02182	0.02058	0.02349	0.01338	0.01219
A9	0.01981	0.01912	0.02222	0.02064	0.02350	0.01349	0.01229
MEAN	0.01897	0.01847	0.02189	0.02081	0.02306	0.01339	0.01216
σ	0.00060	0.00039	0.00024	0.00050	0.00053	0.00006	0.00008
COV	0.03163	0.02112	0.01096	0.02403	0.02298	0.00460	0.00646

Table 3. Natural Frequency Results in Analytical Study

CASE	FREQUENCY	CHANGE	PERCENTAGE
BASELINE	41.625Hz		
DAMAGE CASE 1	41.137Hz	-0.488Hz	-1.172%
DAMAGE CASE 3	40.184Hz	-1.441Hz	-3.462%
DAMAGE CASE 4	41.599Hz	-0.026Hz	-0.063%

Table 4. Natural Frequency Results in Experimental Study

CASE	MEAN	σ	CHANGE	PERCENTAGE
CASE 1	49.8047Hz	0.0000		
DAMAGE CASE 1	48.3235Hz	0.0488	-1.481Hz	-2.974%
CASE 2,3	48.6654Hz	0.0997		
DAMAGE CASE 2	48.9584Hz	0.0977	0.293Hz	0.602%
DAMAGE CASE 3	45.1498Hz	0.0646	-3.516Hz	-7.224%
CASE 4	51.1231	0.0000		
DAMAGE CASE 4	51.4160	0.0000	0.2929Hz	0.573%

In the experimental study, the coherence of all sampling channels is greater than 0.95 within the interested frequency range of 45 Hz to 52 Hz (Fig. 4), which indicates that the signal noise (S/N) ratio is high enough to achieve good estimates of the response. The mode shapes obtained from test results were consistent within the same test case. The small deviation in measured mode shaped demonstrates the accuracy of the test.

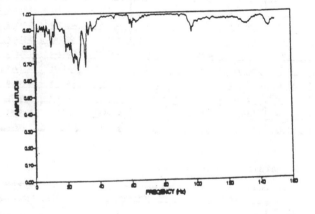

Fig. 4 Typical Coherence Function of Test Response

Results

Modal Damping Ratio

Modal damping ratios obtained from accelerometers associated with the damage cases and related baseline values along with their mean, standard deviation (σ) and coefficient of variation COV are examined. Because of the high COV value, no significant stable changes related to the damage cases can be obtained, suggesting that the traditionally-used damping ratio may not be a good indicator. A comprehensive discussion of damping in structural dynamics may be found in Liang and Lee [1991].

Natural Frequency

Tables 3 and 4 show the analytically and experimentally-obtained natural frequencies of the baseline structure and the damaged structures, respectively. The first mode is the bending mode in the strong axis direction.

The natural frequencies dropped when full depth cracks occurred on the flange (Case 1 and 3), which signifies structural stiffness deterioration. The changes of natural frequency reflect the presence of the damages on the flange. However, very little frequency decrease was noted for Damaged Case 4 with a crack on the web in the analytical study. This very little decrease in frequency is done to a slight change in the moment of inertia of the cross section when the crack was introduced in the web.

For Damage Cases 1 and 3, although the cracks were of the same size, the frequency change due to Damage Case 2 was approximately three times of that of Damaged Case 1 in both the analytical and experimental studies. This indicates that the frequency change in the first mode is more sensitive for cracks developed at the center of the beam than those introduced near the ends. For the same crack length the relative significance of frequency change in a certain mode is determined by the position of the crack. Thus, when the crack occurs closer to the location corresponding to higher relative values of the mode shape, more significant changes of the structural stiffness, resulting in more detectable changes in natural frequency, can be observed.

A comparison between the results of the analytical study and experimental study shows that the frequency changes in the experimental study are larger. The difference is likely to be the result of the finite element approximation and the error of experimental analysis.

Displacement Mode Shape

Displacement mode shapes were examined in both the analytical and experimental studies. Fig. 5 shows the analytical displacement mode shapes corresponding to the baseline, Damage Case 1, and Damage Case 3, respectively. Fig. 6(a) and (b) are the experimental displacement mode shape comparisons of

Damage Case 1, Damage Case 2, and 3 with the corresponding baseline values. Since the displacement mode shape curve of Damage Case 4 is approximately the same as that of the baseline case in both analyses, it is not shown in these Figures. However, in Damage Cases 1 and 3, an increase in the amplitude of displacement mode shape can be observed within a large range of damage locations. This increase in amplitude indicates that flange cracks lead to detectable global changes of the displacement mode shapes.

In order to determine the damage locations from the mode shapes, the differences of the mode shapes for the damage cases with respect to the baseline are shown in Figs. 7 and 8 for the analytical study and experimental study respectively. Because of the slight change observed for Damage Case 4 in the experimental study, this difference curve is not included in Figures 7 and 8.

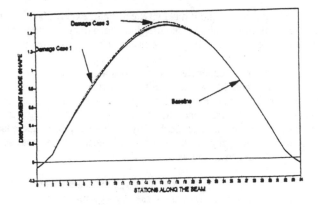

Fig.5 Displacement Mode Shapes in FEM

From both the analytical and the experimental studies, some important observations can be made:

(1) The largest DMS change occurs near the stations where the damages occur.

(2) In the analytical analysis, a comparison of the changes in Damage Cases 1 and 3, the crack at the location close to center of the beam affects the mode shape associated with the first vibration mode more significantly.

(3) The difference curve for Case 2 also clearly indicates the damage location although the amplitude is relatively small (Fig. 8). compared with the results of modal frequency in which no visible changes can be seen, the DMS is more reliable for the "small" damage. Fig. 7 also shows that the change of mode shape is small for the web-damage case when compared to the flange-damaged cases.

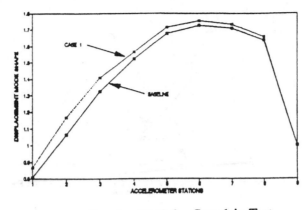

Fig. 6a DMS for Case 1 in Test

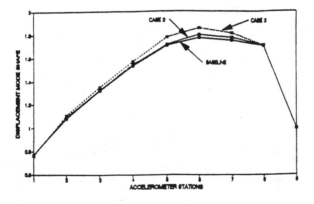

Fig. 6b DMS for Case 2 and 3 in Test

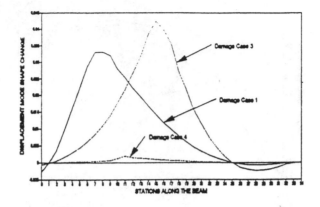

Fig.7 DMS Difference Curves for Cases 1, 3, and 4 in FEM

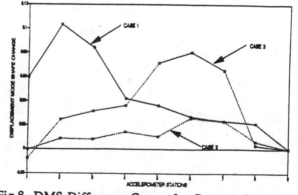

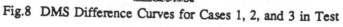

Fig.8   DMS Difference Curves for Cases 1, 2, and 3 in Test

Strain Mode Shape

Fig. 9 shows the analytical strain mode shapes of the baseline value, Damage Cases 1, 3, and 4 respectively, while Fig. 10 shows their differences with respect to the baseline obtained in the analytical study.

In the experimental study, comparisons of strain mode shapes between the damage cases and the relative baseline for Damage Case 1, 2, and 3 are shown in Figs. 11(a) and (b) respectively. Fig. 12 shows their differences as well. Because Damage Case 4 does not affect the moment capacity of the beam significantly, no changes on strain mode shape are measured. Therefore, Damage Case 4 is not shown.

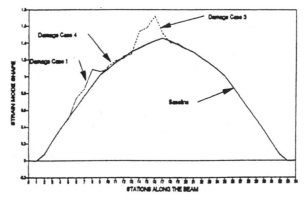

Fig. 9 Strain Mode Shapes for Case 1, 3, and 4 in FEM

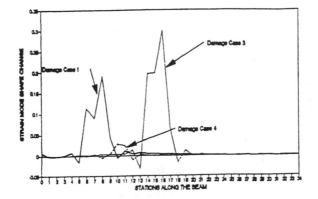

Fig.10 SMS Difference Curves for Case 1, 3, and 4 in FEM

As seen from Figs. 9 through 11, in Damage Cases 1, 2, and 3, increases in the amplitude of SMS is relatively close to the damaged locations. In addition, the SMS shows a much higher sensitivity to the damages as compared with that of DMS because the strain concentration occurs near the cracks. The relatively large localized changes can facilitate the determination of the damage locations. A clear change made by Damage Case 2 is shown at station 6 in the experimental study although there is no significant changes either in frequency or in DMS.

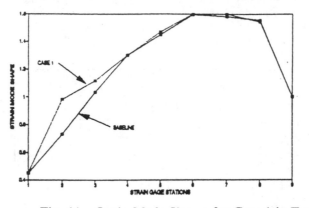

Fig. 11a Strain Mode Shapes for Case 1 in Test

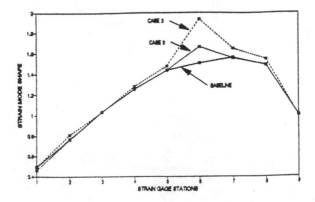

Fig. 11b Strain Mode Shapes for Case 2 and 3 in Test

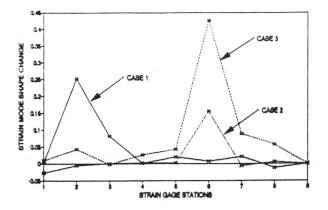

Fig.12 SMS Difference Curves for Case 1, 2, and 3 in Test

From the difference curves, it is also seen that, for the same crack size, the strain mode shape change of Damage Case 3 is more notable than that of Damage Case 1. It is also clearly demonstrated that the significance of strain mode shape change in a given mode is determined by not only the seriousness of the crack but also the position of the crack.

Conclusions

The results of this study showed that: (1) Experimental evaluation of natural frequency is more reliable as compared to the traditionally used damping ratios; (2) Similar cracks at various locations contribute differently to the changes of the modal parameter; (3) The mode having the most significant changes in its parameters is the mode that its DMS takes its largest relative value close to the crack location; (4) Web cracks had insignificant effect on the bending capacity and hence, on the dynamic parameters; (5) SMS proved to be very sensitive in detecting the damaged zone as compared to other modal parameters; (6) Using the changes in modal parameters rather than their absolute values yields significant information about the crack location. Those changes can be used as input to neural networks for on-line damage diagnosis

Acknowledgement

This study is jointly supported by the National Science Foundation of the USA (NO. 150-4642A) and the National Science Council of ROC (NO. 82-0414-P-002-031-BY).

References

Bemasconi, O. and Ewins, D.J. [1989], "Application of Strain Modal Testing to Real Structures", 7th IMAC, Las Vegas, Nevada.

Bishop, R.E.D. and Gladwell, G.M.L. [1963]. "An Investigation into the Theory of Resonance Testing", Proc. Ray. Soc. Phil. Trans. 255(A)241.

Biswas, M, Pandey, A.K. and Sanmman, M.M. [1989]. "Diagnostic Experimental Spectral/Modal Analysis of Highway Bridge," The Intl Journal of Analytical and Experimental Modal Analysis 5(1):33-42.

Clough, R.W. and Penzien, J. [1975]. Dynamics of Structures, MeGraw-Hills, New York.

Ewins, D.J. [1986]. Modal Testing: Theory and Practice, Research Studies Press, England.

Liang, Zhong [1991], Modal Analysis Lecture Notes, Dept. of Civil Engineering, SUNY at Buffalo.

Liang, Zhong and Lee, George C. [1991],"Damping of Structures",Part I, Theory of Complex Damping, National Center for Earthquake Engineering Research Report 91-0004, Oct.

Mazurek, David F. and DeWolf, John T. [1990]. "Experimental Study of Bridge Monitoring Technique", J. of Structural Engineering, V 116 n 9.

Salane and Baldwin [1990], "Identification of Modal Properties of Bridges", J. of Structural Engineering, V 116 n 7.

Yao, G.C., Chang, K.C., and Lee, G.C. [1992]. "Dynamic Damage Diagnosis of A steel Frame", J. Eng. Mech., Vol.118, No.9.

Nondestructive Evaluation With Vibrational Analysis

Robert G. Lauzon, M. ASCE and John T. DeWolf, Fellow ASCE

Abstract

A full-scale highway bridge, in the process of being demolished and replaced, was monitored using vibrational techniques. The bottom of the flange and web of one fascia girder were incrementally cut. Vibrational monitoring during the passage of a test vehicle was used to demonstrate that the vibrational signature of a bridge will change when a major defect occurs. Included in this paper is a discussion of how vibrational studies have been used in the evaluation of bridges.

Introduction

The Federal Highway Administration has reported that there are more than 230,000 deficient or functionally obsolete bridges in the United States. Many of these spans were built in the 1950s and 1960s. Presently, inspections are carried out at approximately two-year intervals, depending on guidelines for the type of bridge and past performance. This is not always sufficient to prevent failures.

At the 1992 Conference on Nondestructive Evaluations of Civil Structures and Materials in Colorado, it was stated that "the present practice of visual inspections at long intervals must be replaced by frequent, automated condition monitoring" and that this should "provide an early warning of distress, support aggressive maintenance programs and promote the timely remedy of emerging deterioration (Working Group on Steel Structures and Materials, 1992).

Vibration monitoring with accelerometers has been used in many areas. Virtually every nuclear power plant, petrochemical plant, and most major manufacturing plants utilize this technology for protecting critical machinery

17

and/or structures. Major airlines utilize the approach to alert pilots of impending danger due to turbine engine bearing or rotor failures. A small number of buildings have been monitored for wind and earthquake forces. The technique has only recently been applied to bridges for monitoring purposes.

The work presented here describes the vibrational monitoring of a full-scale bridge subjected to a destructive test performed by research personnel at the Connecticut Department of Transportation with assistance from researchers at the University of Connecticut.

Literature Review

Most vibration studies of bridges have been one-time studies to determine different vibrational properties, such as natural frequencies, mode shapes or damping ratios. A large number of these have concerned wind induced vibrations, particularly those of suspension bridges. A relatively small number of studies have attempted to apply vibrational information to long-term bridge monitoring. The following reviews applications of vibrational measurements to monitoring bridges.

An extensive field study (DeWolf, Kou and Rose, 1986) of a major four span continuous bridge in Connecticut formed the basis of continued vibrational studies of bridges at the University of Connecticut. The information collected from both traffic and test vehicle induced vibrations, and the knowledge gained on equipment needed for bridge studies, established that vibrational monitoring as used in other fields had application to bridges for the prevention of catastrophic consequences.

A study of the Florida's Sunshine Skyway Cable Stayed Bridge (Jones and Thompson, 1991) was based on obtaining vibrational information for monitoring large bridges. The focus was the behavior under wind, and it was concluded that the work could be continued with a permanent installation to assess the performance of the bridge in storms.

Davis and Paquet (1992) proposed extracting dynamic information from strain measurements for monitoring. Hearn and Ghia, in a report of an ongoing investigation, used dynamic strain records for twenty-nine bridges to detect the free vibration response in order to detect and assess changes in bridge conditions. Again, like the Jones and Thompson study, they established preliminary information as a basis for identifying changes over time.

Other researchers have conducted studies to determine how vibrational properties change when structural deterioration occurs, with applications to bridges. Extensive laboratory tests of bridge models with moving loads were conducted at The University of Connecticut (DeWolf, Lauzon and Mazurek, 1988; Mazurek and DeWolf, 1990). It was demonstrated that major deterioration is detectable, noting that some of the lower natural frequencies and mode shapes change with cracking, support displacements and connection problems. They were also able to correlate experimental data with finite element analyses. Mazurek (Mazurek, Jordan, Palazzetti and Roberts, 1992), in efforts based on work at the University of Connecticut, determined frequency and mode shape changes as a defect propagated in a beam. The results were analytically correlated with the placement of a hinge and rotational spring in an equivalent beam.

Spyrakos, Chen, Stephens and Govindaraj (1990) used test beams to evaluate the threshold for which damage can be detected using the dynamic characteristics. They found a correlation between the reduction in natural frequency with an increase in damage. Agbabian, Masri, Traina and Waqfi (1990) used a laboratory bridge model to study the potential of using the dynamic response to detect structural changes. Turner (1990) performed experimental studies on a simply supported beam to determine natural frequency shifts due to defects. He concluded that natural frequency changes may be used as indicators to detect structural damage. He demonstrated the viability of measuring natural frequencies from traffic induced vibrations, based on seven full-scale bridge tests. Tang (1989) used finite element models to determine how natural frequencies and mode shapes change due to structural imperfections. He noted that the mode shapes are a good indicator for detecting damage in a bridge's superstructure and proposed using this work as a basis for assessment of bridges.

Hearn and Testa (1991) in related work demonstrated that vibrations can be used for nondestructive inspection of structures. They found that the natural frequencies and modal damping coefficients change with structural deterioration, based on experiments with welded steel frames and wire rope.

Modal analysis is based on using an impact device or known loading, and then making comparisons of the data at the point of interest with the impact or loading data.

A recent study of older bridges in Turkey (Uzgider, Sauli, Caglayan and Piroglu, 1992)) was based on using test vehicles to conduct modal analysis. The vehicles were outfitted with accelerometers for determination of the input data. The collected data will be used as a base for later comparisons.

Raghavendrachar and Aktan (1992) used the modal technique to review a reinforced concrete bridge. They concluded that vibrational studies can provide information useful in monitoring. They noted the need for additional research to perfect the approach.

Modal analysis of bridges, has not always been successful because of the difficulty in making comparisons between the input and measured data. Unlike many model studies, bridges are neither homogeneous nor are all members fully connected together to behave as an integral unit. Additionally, the modal analysis method requires closing a bridge to conduct the monitoring with a test vehicle or other input device.

An alternative vibrational analysis approach is to use the ambient method. This can be based on random loading, such as that from traffic, and it involves making comparisons of the data from different sensors, as opposed to making comparisons of the input and response data as used in modal analysis. Thus, in the ambient method it is not necessary to know the input functions. All that is needed is the use of multiple sensors, with simultaneous measurements, so that responses from different sensors can be compared.

Reed and Cole (1976) proposed what they called the "random-dec technique" for use in monitoring bridges. This technique is based on measurement of ambient vibrations. They were able to detect crack growth with sensors placed in close proximity to the crack. They concluded that traffic induced excitations can be used to develop a practical method for monitoring the structural integrity of bridges. Recommendations involving major additional studies were presented in their report to fully develop the method.

Musa (1990) determined dynamic properties for a bridge subjected to truck traffic using an approach based on ambient monitoring. Included were the natural frequencies, mode shapes and damping ratios. He found that field data compared well with finite element analyses.

Alampalli, Fu, and Aziz (1992) placed a vibrational monitory system on a bridge in New York. The system used traffic induced vibrations. They concluded that frequencies, mode shapes, and comparisons of other data can be used with consistency for identifying modal parameters for monitoring.

The previously noted vibrational studies at The University of Connecticut have been based on the ambient vibration technique. The results were used to work with Vibra·Metrics in Hamden, Connecticut, to design and build a prototype vibrational monitoring system for bridges. The system was used for data collection on a Connecticut bridge during a period of one and a half years (O'Leary, Bagdasarian and DeWolf, 1992). The researchers were able to: (1) extract meaningful frequency spectra from normal traffic flow; (2) directly determine the natural frequencies and mode shapes; (3) demonstrate that this vibrational information is stable and repeatable; (4) compare natural frequencies and mode shapes from the field data to finite element analysis results; (5) collect repeatable data during all possible weather conditions.

Little has been done to introduce structural deficiencies in actual bridges and determine how changes in the structural integrity alter the global properties. Biswas, Pandey and Samman (1990) loosened bolts in a connection for a two-span continuous bridge, noting some change in the vibrational information.

The following presents initial results from a study in Connecticut in which a crack was introduced in a bridge.

Bridge Test Procedure

The structure utilized for this test was a three-span simply-supported bridge (approx. 180-ft in total length) which consisted of eight simply-supported rolled beams (W36x165) across the width of the bridge which supported an 8.5-in. cast-in-place concrete deck. The structure is shown in Fig. 1. Diaphragms (W30x99) were located at the midspan, quarter·span and at the ends of the girders and were attached to the webs of the girders using riveted angles. The bridge structure was replaced in three stages. Before the destruction phase of Stage Three and immediately following the completion of Stage Two, the destructive test took place. The cross section of the portion of the western end span used for this study is shown in Fig. 2.

Equipment

Accelerometers were attached to the underside of the girders using magnets designed and connected to an eight-channel digital tape recorder using cables. The tape recorder was battery-powered and utilized a Digital Audio Tape that could retain one hour of real-time signals produced by the accelerometers. Following collection of the data, the recorder was brought into the laboratory where it was combined with a Digital/Analog Converter to transfer the data to Personal Computer (PC). After transfer to the PC, the data was analyzed using signal-processing software.

Test Procedure

To approximate a cracked girder, the exterior girder was cut incrementally with a torch. This girder ("C") was selected for cutting due to the higher probability that such a girder would first develop a crack on an in-service bridge. Due to the girder's weakened condition, a temporary support structure was provided under the test span so that any unanticipated movement of the structure would be arrested before it could become a hazard to the test vehicle. Static deflection of the bridge was measured after each cut and dynamic deflection was observed to ensure no contact was made with the support during the test.

A full-size pickup truck weighing approximately 5400 pounds was used as the test vehicle. Based on previous studies, it is expected that normal traffic should consistently excite the lowest 2 to 6 natural frequencies, including both flexural and torsional mode shapes. Above 100 Hz, little information is available.

Eight accelerometers were used for the primary test and were located on each of the three girders of the test span. The locations were chosen to provide a complete description of the vibration of the bridge. Using the data from the accelerometers at these locations, a comprehensive vibrational signature can be established. The crack approximation was located near the midspan of Girder "C". The cut was not made directly at midspan due to the existence of a diaphragm at that location. Before any cuts were made, the vibrational signature for the unaltered test span was fully defined. This was done using a number of tests with the vehicle crossing the span. The vehicle started from rest, off the bridge, and accelerated across the test span. Three sets of vehicle passes were done with the unaltered structure. A minimum of 12 passes were made for each set so that an average frequency spectrum could be established. The cut was incremented, beginning with the lower flange. It was then extended into the web in increments, and a full set of vehicle passes made between each cut. Following the final set of vehicle passes with the cut extended 24-in. up from the flange, the testing was terminated without incident.

Test Results

Overall, the static deflection of the test span increased very slightly from the unaltered condition and did not result in the span coming in contact with the temporary support at any time during the test.

The combination of the resonant frequencies, and their respective amplitudes and corresponding mode shapes can be termed a "vibrational signature" for a structure. Vibrational monitoring is based on the premise that a change in the stiffness will cause a change in the vibrational signature. Theoretically, the resonant frequencies will shift their location due to a reduction in structural stiffness, the mode shapes will change, and the respective amplitudes will be affected.

Figure 3 shows a set of typical frequency spectra from the vibrational analysis of the structure. Included are the spectrum prior to cracking and those with incremental cracking. The peaks in these spectra correspond to the resonant or natural frequencies of the structure. These frequencies are a function of the stiffness and mass of the structure. Each resonant frequency is associated with a single mode shape, which is the deflected shape the structure would take if it vibrated only at that frequency.

Table 1 contains the frequencies attained at each of the six stages of the test for two accelerometer locations. Channel 4 was located at the midspan of the girder which was cut, and Channel 6 at the midspan of the opposite fascia girder. The frequency values (Freq.) represent vibrational frequencies that were detected at the respective accelerometer locations during that stage of the test. The amplitude (Amp.) values represent the amplitudes of the corresponding frequencies in millivolts.

The amplitudes of each spectra were normalized with respect to the maximum amplitude within that spectrum. It was anticipated that normalization would facilitate the identification of the frequencies that shifted and by what degree they shifted. The resolution of the results is 0.2 Hz and is governed by the length of time from start to finish of the vibrational response from the passes of the test vehicle.

The table demonstrates that the fundamental frequency remained fairly constant while other frequencies, as prominent as the fundamental, appeared during the latter stages of the test. Using a threshold, arbitrarily set at 60 percent of the maximum amplitude within a spectrum, it was anticipated that the appearance of additional frequencies could be better determined and quantified. Therefore, the amplitude values shown in parentheses fall below the threshold for that particular frequency spectrum.

The maximum amplitudes occurred mostly in the final stage when the cut was extended 24-in. into the web and

were associated with the fundamental frequency. This was
expected given that the reduced stiffness would produce a
more   flexible   structure,   thereby   producing   larger
amplitudes of vibration.   Channel 6 at midspan, furthest
away from the parapet, recorded the largest amplitudes of
all the channels during each stage of the test.

Midspan - Fascia Girder near Crack (Channel 4)   This
channel represents the accelerometer closest to the cut.
The results show a fundamental frequency in the 8.2 to 8.6
Hz range.   The fundamental frequency falls below the
threshold immediately at the first alteration.   An
additional frequency develops as the crack begins in the
web. The final alterations produce added frequencies above
the threshold at this level of the cut which was also
evident with other channels.

Midspan - Fascia Girder Opposite Crack (Channel 6)
Added frequencies are seen only in alterations 4 and 5, and
like Channel 4, a transfer in the dominant frequency from
the second (torsional) to the fundamental (bending) became
apparent as the test progressed.   As was the case with
Channel 4, the frequency at 12.6 Hz (torsional) was
replaced with ones at 12.2 and 13.4 Hz with alterations 4
and 5.

Conclusions

    The immediate effect of the crack approximation was
seen in the results for some channels before the structural
integrity of the test span was compromised.

    A change in the frequency spectra with the addition of
prominent   new   frequencies   is   evident   as   the   crack
approximation progressed and is substantial enough to show
structural change.

    With the resolution of the experimental data, a drop
in the fundamental frequency from approximately 8.6 Hz to
approximately 8.2 Hz with the first alteration is not
substantial enough to determine structural change without
long-term monitoring of the span to determine the stability
of the fundamental frequency.

    The transfer in dominant frequency from the second
resonant frequency (torsional) to the fundamental frequency
(bending) is consistent for each of the channels monitored
and could be useful in determining structural change.

    A decrease in frequency amplitude indicates a likely
change in mode shape for the fundamental frequency as the
test progressed. This identifies the use of mode shapes as
a portion of the vibrational signature that could indicate

structural change.

Bridge monitoring will only be successful if it incorporates a review of many different variables, none of which would be adequate alone in predicting structural integrity problems. The attempt to use only limited variables is why earlier studies have not successfully demonstrated that vibrational monitoring is applicable to bridges.

Bibliography

Agbabian, M.S., Masri, S.F., Traina, M.I., and Waqfi, O., "Detection of Structural Changes in a Bridge Model," Structures Congress Abstracts, ASCE, April 30 - May 3, 1990, p. 111.

Alampalli, S., Fu, G., and Aziz, I.A., "Nondestructive Evaluation of Highway Bridges by Dynamic Monitoring," Proceedings of Conference on "Nondestructive Evaluation of Civil Structures and Materials," University of Colorado, Boulder, Colorado, May 1992.

Biswas, M., Pandey, A.K. and Samman, M.M., "Diagnostic Experimental Spectral/Modal Analysis of a Highway Bridge," The International Journal of Analytical and Experimental Modal Analysis, Vol. 5, No. 1, January 1990, pp. 33-42.

Davis, A.G. and Paquet, J., "Monitoring Bridge Performance Using NDE Techniques," Proceedings of Conference on "Nondestructive Evaluation of Civil Structures and Materials," University of Colorado, Boulder, Colorado, May 1992.

DeWolf, J.T., Kou, J.W. and Rose, A.T., "Field Study of Vibrations in a Continuous Bridge," Third International Bridge Conference, Pittsburgh, Pennsylvania, June 1986, pp. 103-109.

DeWolf, J.T., Lauzon, R.G. and Mazurek, D.F., "Development of a Bridge Monitoring Technique," Proceedings: Bridge Research in Progress, Iowa State University, Ames, IA, September 1988, pp. 65-68.

Hearn, G. and Ghia, R., "Response-Based Structural Condition Monitoring," Proceedings of Conference on "Nondestructive Evaluation of Civil Structures and Materials," University of Colorado, Boulder, Colorado, May 1992.

Hearn, G. and Testa, R.B., "Modal Analysis for Damage Detection in Structures," Journal of the Structural

Division, ASCE, Vol. 117, No. 10, 1991, pp. 3042-3063.

Jones, N.P. and Thompson, J.M., "Ambient Vibration Survey and Preliminary Dynamic Analysis: Sunshine Skyway Cable-Stayed Bridge," Department of Civil Engineering, The Johns Hopkins University, Baltimore, Maryland, 1992.

Mazurek, D.F. and DeWolf, J.T., "Experimental Study of Bridge Monitoring Technique," Journal of Structural Engineering, ASCE, Vol. 116, No. 9, September 1990, pp. 2532-2549.

Mazurek, D.F., Jordan, S.R., Palazzetti, D.J., Robertson, G.S., "Damage Detectability in Bridge Structures by Vibrational Analysis," Proceedings of Conference on "Nondestructive Evaluation of Civil Structures and Materials," University of Colorado, Boulder, Colorado, May 1992.

Musa, S.J., "Determination of the Dynamic Properties of an In-Situ Bridge from Response Data," Dissertation, 1990.

O'Leary, P.N., Bagdasarian, D.A. and DeWolf, J.T., "Bridge Condition Assessment Using Signatures," Proceedings of Conference on "Nondestructive Evaluation of Civil Structures and Materials," University of Colorado, Boulder, Colorado, May 1992.

Raghavendracher, M. and Aktan, A.E., "Flexibility by Multireference Impact Testing for Bridge Diagnostics," Journal of Structural Engineering, ASCE, Vol. 118, No. 8, August 1992, pp. 2186-2203.

Reed, R.E. and Cole, H.A., "Mathematical Background and Application to Detection of Structural Deterioration in Bridges," NASA Technical Report FHWA-RD-76-181, 1976.

Spyrakos, C., Chen, H.L., Stephens, J., and Govindaraj, V., "Evaluating Structural Deterioration Using Dynamic Response Characterization," Proceedings Intelligent Structures, Applied Mechanics, 1990, pp. 137-154.

Tang, Jhy-Pyng, "Vibration Measurement and Safety Assessment of Bridges," The Engineering Index Annual, 1989, p. 1051.

Turner, J.D., "An Experimental and Theoretical Study of Dynamic Methods of Bridge Condition Monitoring," Dissertation, 1990.

Uzgider, E., Sauli, A.K., Caglayan, O., and Piroglu, F., "Full Scale Static Testing of Bridges Using Tiltmeters," Fourth International Conference on Structural Failure,

Product Liability and Technical Insurance, Technical University of Vienna, Vienna, Austria, 1992.

Working Group on Steel Structures and Materials, Proceedings of Conference on "Nondestructive Evaluation of Civil Structures and Materials," University of Colorado, Boulder, Colorado, May 1992.

| Channel | Unaltered | | Flange Cut | | Altered Conditions | | | | | | | |
| | | | | | 6" into Web | | 12" into Web | | 18" into Web | | 24" into Web | |
	Freq.	Amp.	Freq.	Amp.	Freq.	Amp.	Freq.	Amp.	Freq.	Amp.	Freq.	Amp.
4	8.6	.02137	(8.4)	.01439	8.2	.03294	8.2	.03418	8.4	.02516	8.4	.00338
	12.6	.03303	12.6	.03110	11.4	.02530	12.6	.03485	11.2	.02296	(9.8)	.00172
	—	—	—	—	12.6	.0378	—	—	12.2	.0360	11.8	.00306
	—	—	—	—	—	—	—	—	13.4	.0367	13.4	.00215
6	8.6	.0268	8.6	.01936	8.2	.04304	8.4	.0378	8.4	.0327	8.2	.0459
	12.6	.0296	12.6	.03205	12.6	.04505	12.6	.0275	12.2	.0441	12.0	.0357
	—	—	—	—	—	—	—	—	13.4	.0337	13.4	.0302
	—	—	—	—	—	—	—	—	—	—	—	—

Note: Frequencies with amplitudes less than 60% of the maximum frequency amplitude for that channel in that test stage are shown in parentheses.

Table 1. Amplitude and Value of Frequencies obtained during vibrational monitoring of bridge superstructure.

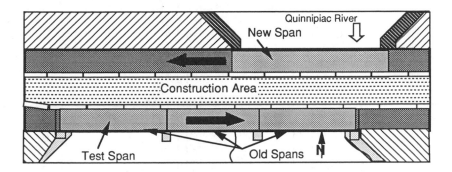

Figure 1. Plan View of Bridge Site

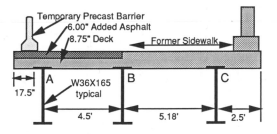

Figure 2. Cross Section of Test Span

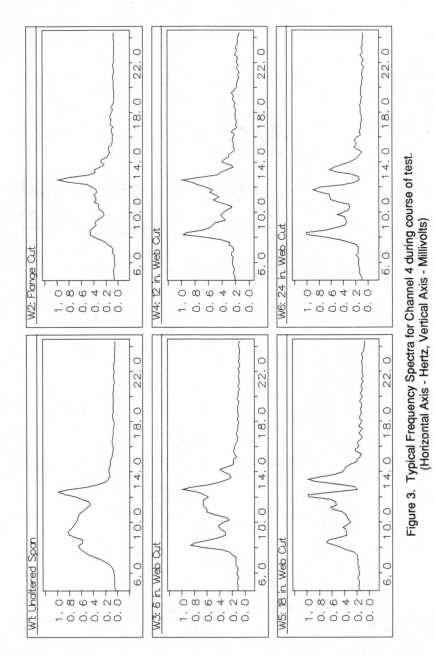

Figure 3. Typical Frequency Spectra for Channel 4 during course of test.
(Horizontal Axis - Hertz, Vertical Axis - Millivolts)

MAGNETIC FLUX LEAKAGE FOR BRIDGE INSPECTION

Charles H. McGogney, P.E. *

Abstract:

For many years the Federal Highway Administration and
before that the Bureau of Public Roads had focused
its nondestructive evaluation program on development
of better tools to inspect components of steel
bridges. In the early seventies the emphasis was
changed and attention was given to the need for an
inspection device that would detect and assess the
condition of the steel elements of concrete bridge
structures. This gave rise to the development of the
Magnetic Field Disturbance (MFD) method and the
ensuing development of the Magnetic Perturbation for
Cables (MPC) inspection system. At present the
Magnetic Flux Leakage Inspection System (MFLIS) for
both main cables of suspension bridges and steel
elements of concrete structures is under development.
This paper summarizes the development of the
aforementioned systems and field and laboratory
applications.

Introduction:

Of the many problems confronting the State bridge
highway departments perhaps the most perplexing is
the corrosion of steel. It is even more of a problem
when the steel elements are embedded in concrete
beams and boxes or comprise the main cables of
suspension bridges. The bridge designers and
engineers go to great lengths to protect the steel
from corrosive environment,

* Metallurgist, Federal Highway Administration,
Office of Advanced Research, Physical Research
Division, 6300 Georgetown Pike, McLean, VA 22101

however, corrosion due to stress and/or fatigue does
occur and poses a serious problem for the maintenance
engineers. It may be that only a small amount of rust
is the only visible sign of distress yet this
condition left uncorrected, could be catastrophic.
To address this problem, the Federal Highway
Administration (FHWA) Office of Research and
Development, over the past twelve years, has
sponsored research to develop a nondestructive
inspection system that can detect and evaluate the
presence of corrosion in steel elements of reinforced
concrete and main cables of suspension bridges.

Background:

As far back as the mid seventies, FHWA recognized the
need for a nondestructive test method to detect
deterioration of reinforcing steel in concrete boxes
and beams of highway bridges. Following the
competitive bid process a research contract was
awarded, to address this problem and develop a
cursory scanning system that the State highway
departments could use for bridge inspection. After a
careful evaluation of 15 candidate methods it was
decided that the magnetic flux leakage method was
most likely to succeed in fulfilling this need. [4]

The magnetic flux leakage technology has been around
for many years. However, field implementation must
be carefully controlled and the signature
characteristics understood. As an example, take a
spherical volume anomaly with a magnetic permeability
(u') embedded in a ferromagnetic material (steel)
with a permeability (u) and a magnetic field (Ho)
applied along the X direction as shown in Figure 1.

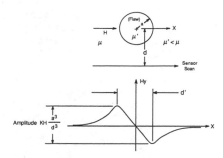

SIGNATURE FEATURES:
  1. Bipolar and symmetric
  2. Polarity as shown for fracture gap or corrosion pit; polarity
     reversed for soft spot in steel of "ferrule" splice
  3. Peak separation d' equals depth to flaw d (independent of size)
  4. Amplitude indicates flaw volume (dependent on depth)
  5. Maximum amplitude obtained when scan path is directly over flaw
  6. Amplitude decreases with flaw depth d (decrease: $1/d^3$)
  7. Signal amplitude increases continuously as the magnetic field (H) is increased

Figure 1. Sketch of spherical void in an infinite
magnetized ferromagnetic matrix and the resulting
field perturbation in that direction

The magnetic field component in the Y direction (Hy)
is sensed and a continuous plot produces the record
or signature.

Equipment Development:

The first generation MFD consisted of a stationary DC
electromagnet with a Hall Effect Sensors located
between pole pieces to measure the disturbances
produced by the anomaly. An analog strip chart
recorder was used to record the signature. The
success of these early stationary trials led to the
construction of an inspection cart to support the
magnet and sensor unit. The cart was equipped with
four wheels to ride on rails that were supported by
hangers that were designed to hang from the bottom
flange of a typical Texas type "C" reinforced
concrete I beam. After further laboratory trials
field testing of the MFD on an in-service viaduct was
conducted to evaluate the total system operation.
The longitudinal scanning and lateral indexing of the
magnet/sensor unit are encoded and controlled by a
remote module. Unexpectedly, signatures from these
early scans were complicated by the response from
stirrups, chairs, and artifacts that were part of the
beam construction. At this point some limited
analyses using subtraction procedures showed some
promise in suppressing the undesirable features.

Under a new contract a second generation of the MFD was pursued addressing the needs for improvement that had become obvious in the earlier field trial. [5] A microprocessor was incorporated in the system to digitize the storage of data and enhance the signal analysis as shown in Figure 2.

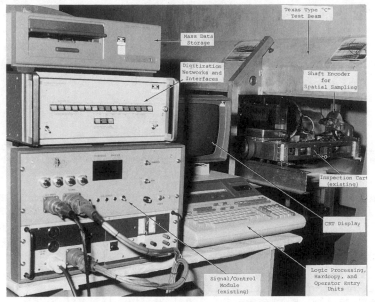

Figure 2. Photograph of signal processing units with magnetic field disturbance system in the laboratory

In addition to the subtraction procedure a correlation method of analysis was developed, using the microprocessor to distinguish flaw signals from other signals. Mathematical flaw signal developments (algorithms) were developed and the recorded data plotted as a function of the scan length. This information was presented on a monitor in almost real time and hard copy printout was possible when needed as shown in Figure 3.

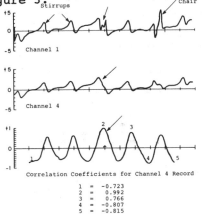

Correlation Coefficients for Channel 4 Record

1	=	-0.723
2	=	0.992
3	=	0.766
4	=	-0.807
5	=	-0.815

Note: Flaw is 1/2-in. separation in 1-3/8 in. diameter bar (1 in. = 2.54 cm)

Figure 3. Reconstructed display of digitized magnetic signatures before and after enhancement and correlation for flaw in 1 3/8" ∅ bar.

A second field trial was conducted on the same
viaduct as in the first field trial and in some cases
on identical post-tensioned I beams. In one case a
1 3/8" diameter, 60 foot long, steel rod had
fractured and "fired out" of the aluminum duct of the
beam. This left an open duct that made it possible
for the inspection team to insert, back into the open
duct a steel rod with known anomalies of various
lengths to simulate breaks and separations at known
locations as shown in Figure 4.

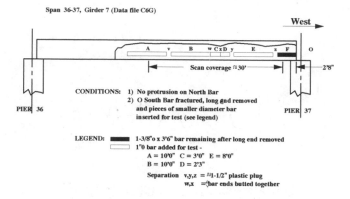

Figure 4. Sketch of concrete beam with open duct
where a 1-inch diameter steel rod with known
separation was inserted and inspected.

Subsequent scanning of this beam with the MFD showed
excellent correlation between known anomalies their
size and locations with that of the MFD signal
magnitude and location on the recordings as shown in
Figures 5 and 6.

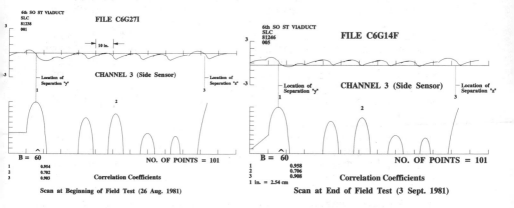

Figure 5. Scan at
beginning of field
test (26 Aug. 1981)

Figure 6. Scan
at end of field
test (3 Sept. 1981)

Another excellent opportunity occurred in 1984,
involving a full size prestressed concrete beam/deck
configuration that was being fatigued tested at
Ferguson Structural Engineering Laboratory (FSEL) at
the University of Texas at Austin.[2] The MFD system
was moved to FSEL and set up under the concrete beam.
At specific intervals in the fatigue test the MFD
scanned the length of the beam and any significant
signals received at given locations were marked at
that location on the beam along with the number of
fatigue cycles and the date. After completion of the
test, excavations were made at the locations that the
MFD unit had identified as possible areas where the
wire strands had fractured. In all cases where two
or more individual wires had fractured in the seven
wire strands excellent correlation between actual
fracture and MFD signature were obtained. More
recent work involving the MFD was conducted at the
University of Wisconsin at Milwaukee. [3] The purpose
of this study was to evaluate the performance of the
MFD both in the laboratory and field and to make
recommendations for further improvement of the

system.   Three primary areas in need of improvement
were identified.   They are the size and weight of the
scanning cart and system components, the outdated
electronics and present flaw discrimination methods.

In 1987 based on the experience gained with the MFD
system and after a thorough review of the inspection
methods available, FHWA chose to further pursue the
magnetic flux leakage technology for development of a
nondestructive cursory scanning inspection system for
detection of anomalies in main cables of suspension
bridges.

The rudimentary prototype developed is known as the
Magnetic Perturbation for Cables (MPC) inspection
system. [1]   The system incorporates an electromagnet
with 12 Hall Effect Sensors connected in pairs to
form 6 data channels that move as a unit along the
cable under inspection.   The output of the sensors is
recorded and displayed to assess flaws (breaks) in
the steel elements.   After several modifications to
the system, the working prototype MPC is an
integrated electromechanical system that consists of:

- a magnetic sensor assembly attached to a
  moving electromagnet

- electrical and hydraulic motors and
  actuators

- a electronic control and data acquisition
  system directed by a remotely located host
  computer, and

- a self-propelled frame housing for the
  inspection components (see Figure 7).

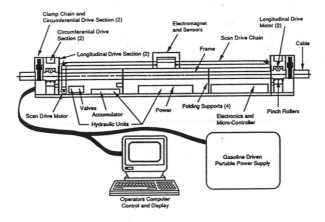

Figure 7. Block diagram of Magnetic Perturbation System for Cable Inspection

The system is designed to be placed directly upon a bridge cable and is capable of moving itself longitudinally and circumferentially about the cable at approximately 1 inch per second acquiring perturbation data, at precise 0.1 inch intervals. The entire unit weighs approximately 2800 pounds and is 17 feet long. The MPC has been successfully demonstrated (tested) on main cables of five suspension bridges and on two occasions on cable stay specimens that were subject to fatigue and proof loading in the FSEL at the University of Texas at Austin (see Figure 8).

Figure 8.   Picture of Magnetic Perturbation for Cable Inspection System on typical suspension cable of bridge

Considering the success of both the MFD and the MPC systems it behooved the FHWA to further pursue research in magnetic flux leakage technology mindful of the need to lighten the weight and shorten the length of the systems and also to improve the flexibility to scan over cable hangers, clamps, and other details on the cables.

After award of a competitive bid contract, a feasibility study was initiated to submit a design concept for improving the prototype MPC and also that of the MFD. FHWA has accepted the design concept submitted by the contractor and is now sponsoring the development. The design calls for the use of permanent magnets as opposed to the electromagnets used in the MPC and the MFD units. The new system will be known as the Magnetic Flux Leakage Inspection System (MFLIS) for both bridge cables and concrete

box and beam inspection. The system will be both[6]
lighter in weight and easier to use. A modular
approach is being pursed whereby permanent
magnet/sensor modules can be interchanged among
various carrier/transport mechanisms. Unlike the MPC
the MFLIS will entirely surround the cable with
magnet/sensor modules providing complete
circumferential inspection in a single pass and
articulate during transit to pass over cable bands on
suspension bridges. The design also calls for
inspection of an entire steel reinforced beam in a
single pass.

The magnet design is based on a magnetic "core
assembly" of two transversely magnetized NbFe B35
magnets, with a low carbon steel space between them
and a low carbon steel pole stub extending at each
end as shown in Figure 9.

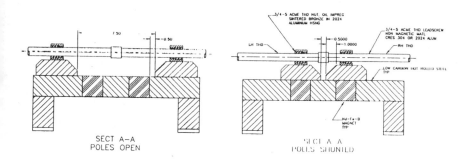

Figure 9. Magnetic and pole pieces in open and
shunted positions

The reason for using two magnets in the core assembly is to facilitate shunting. Magnetic shunting is accomplished by two low carbon steel sections which are moved inward from an "open" position over the pole stub extensions, to a shunt position where they are positioned above the magnets. The shunts are moved on non-magnetic linear rolling elements by a coupled pair of RH/LH threaded nonmagnetic lead screws which are driven by a reversible gearmotor. Each complete magnet module, including the shunt motor, will be about 25 inches long and weigh about 80 pounds. Based on previous experience, suppliers of the NdFe B35 magnet material expect the magnet module to generate field equivalent to a 40,000 ampere-turn electromagnet.

DISCUSSION:

Implementation of the magnetic flux leakage
technology for detecting anomalies in steel elements
of boxes and beams of concrete bridges and main cable
of suspension bridges shows considerable promise.
However, there are a number of deficiencies that
exist within the existing developments that need to
be addressed.

- The electronic hardware has become outdated
  and needs to be updated.

- The correlation analysis needs to be
  improved to speed up interpretation of the
  data.

- The size and weight of the system needs to
  be reduced.

The present ongoing contract the FHWA is conducting
does address these deficiencies and intends to
develop a reliable, practical, and economical
nondestructive inspection system for detection and
assessment of anomalies in steel elements of concrete
and main cable of suspension bridges.

REFERENCES:

1. Barton, John R; Cecil M. Teller, and Sidney A. Suhler, *"Design, Develop and Fabricate a Prototype Nondestructive Inspection and Monitoring System for Structural Cables and Strands of Suspension Bridges: Volume 1, Final Report"*, Report FHWA/RD-89-158, May 1989

2. Beissner, R. E. and J. R. Barton, *"Laboratory Test of Magnetic Field Disturbance (MFD) System for Detection of Flaws in Reinforcing Steel in Prestressed Concrete Bridge Members"*, Contract DTFH61-83-C-00090, U. S. Department of Transportation, 1984

3. Ghorhanpoor, A, G. R. Steber, and T. E. Shew, *"Evaluation of Steel in Concrete Bridges; the Magnetic Field Disturbance (MFD) System"*, Report FHWA-SA-91-026, U. S. Department of Transportation, May 1991.

4. Kusenberger, F.N. and J.R. Barton, *"Detection of Flaws in Reinforcing Steel in Prestressed Concrete Bridge Members."*, Report FHWA/RD-81/087. U. S. Department of Transportation, 1981

5. Kusenberger, F. N. and R. S. Birkelbach, *"Detection of Flaws in Reinforcing Steel Concrete Bridge Members"*. Report FHWA/RD-83/081, U. S. Department of Transportation, 1983

6. Matzkanin, George A., Thomas Stephens and Anthony Chalkley, *"Design Report - Phase I, Magnetic Flux Leakage Inspection System (MFLIS), Final Report"*, Contract DTFH61-91-C-00039, U. S. Department of Transportation, January 1992.

Signal Analysis for Quantitative AE Testing

E. N. Landis[1] and S. P. Shah[2]

Abstract

Quantitative acoustic emission (AE) techniques can be used to monitor crack growth and to deduce microfracture mechanisms in quasi-brittle materials. Multichannel data acquisition systems are used to record the AE waveforms. The results of a quantitative AE analysis are quite sensitive to the interpretation of the recorded waveforms. Several techniques are presented here which facilitate AE data interpretation. Included are noise reduction and signal enhancement methods, and time domain deconvolution methods which are used to recover microcrack fracture parameters.

Introduction

An acoustic emission (AE) is the spontaneous release of localized strain energy in a stressed material. This energy release causes the propagation of stress waves which can be detected at the surface of the material. Quantitative AE techniques can be used to monitor crack growth and to deduce microfracture mechanisms in quasi-brittle materials. This information is obtained through the analysis of surface displacement transients in a loaded specimen, and through the application of seismological inverse methods. Quantitative AE techniques have been successfully applied to a variety of materials (Enoki and Kishi 1988, Ohtsu et al. 1989, Ouyang et al. 1991). A typical AE test setup is shown in Figure 1.

Quantitative AE theory is based on the analysis of transient displacement data recorded at different locations on a test specimen. These records are the output of high speed data acquisition equipment. In order to maximize the amount of

---

[1] Student Member, ASCE; Research Assistant
[2] Member, ASCE; Director, NSF Center for Advanced Cement-Based Materials, Northwestern University, Evanston, Illinois, 60208, USA

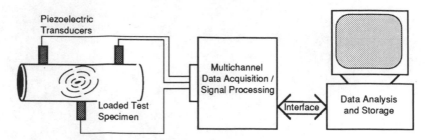

Figure 1. Typical Acoustic Emission Test Setup

information that can be extracted from these data records a variety of signal analysis routines are employed. The analysis routines described in this paper fall into two broad categories: filtering routines for signal enhancement and noise reduction, and deconvolution routines required for quantitative AE inversion. This paper presents a variety of routines which were found to be very useful in the analysis of AE data. The work described is a part of an ongoing experimental research program on fracture mechanisms of cement-based materials.

Quantitative AE Background

Source Location

The location of an acoustic emission source is of primary importance for investigations of damage localization as well as the detection of individual material flaws. The source location is evaluated through the differences in P-wave arrival times at the different transducers in an experimental array. This is analgous to the location of an earthquake epicenter in seismological analysis. The location is determined by solving the following equation:

$$\Delta t_{a-b} = \frac{\left(|x_a - x| - |x_b - x|\right)}{c_p} \tag{1}$$

where x is the unknown location of the AE event, x_a and x_b are the locations of the ath and bth transducers, respectively, c_p is the known velocity of the P-wave in the medium, and Δt_{a-b} is the difference in signal arrival times at the ath and bth transducers. Since this is a system of nonlinear equations, an iterative solution is often most effective. In three dimensions a minimum of four noncoplaner transducers are required to solve the problem uniquely. Additional transducers can improve the solution since the effect of random measurement errors is reduced. A least squares approach may be used.

The accuracy of the source location is related to a number of factors such as transducer array geometry and P-wave velocity measurement, but the accuracy is most closely linked to the quality of the signal arrival time measurments. It this motivation for quality arrival time measurements that certain signal processing routines were developed. These routines are discussed in detail below.

Source Characterization

The characteristics of the acoustic emission source can be deduced through what has often been referred to as quantitative acoustic emission analysis (Hsu et al. 1977). In this analysis, all aspects of the AE process are evaluated. These processes include the source event, the resulting wave propagation through the medium, and the measurement of the waveform at the specimen surface by the AE transducer. The characterization is made by "removing" the effects of the wave propagation and measurement processes.

The basis of quantitative acoustic emission analysis is an integral solution to the differential equation of motion. If body forces, surface tranctions, and initial surface displacements are neglected, then the solution may be written:

$$u_i(t) = G_{ij,k}(t) * D_{jk}(t) \qquad (2)$$

where '*' denotes a convolution integral. Here $D_{jk}(t)$ is a tensor representing the AE source (e.g. microcrack), $u_i(t)$ is the displacement transient at the surface of the specimen, $G_{ij,k}(t)$ is the elastodynamic Green's function which represents the propagation of the stress wave from the source to the receiver (Aki and Richards 1980). The effects of the transducers and measurement system are taken into account by the convolution of equation (2) with an appropriate system response function, $R_i(t)$, such that:

$$V(t) = R_i(t) * \left\{ G_{ij,k}(t) * D_{jk}(t) \right\} \qquad (3)$$

where $V(t)$ is the measured voltage transient. The goal of the quantitative AE analysis is to deduce the source characteristics from the measured signals. This can be accomplished through the inversion (deconvolution) of equation (3). If the system response function and the elastodynamic Green's function are known prior to the experiment, then the source function is evaluated through the following equation:

$$D_{jk}(t) = V(t) * \left\{ G_{ij,k}(t) * R_i(t) \right\}^{-1} \qquad (4)$$

Using this approach, an AE source is characterized according to its volume, fracture mode and crack plane orientation. All of these characteristics can be estimated from the source function $D_{jk}(t)$ (Enoki and Kishi 1988).

There are a number of different methods which can be used to evaluate equation (4) (e.g. Scruby et al. 1986, Enoki and Kishi 1988, Landis and Shah 1993). All methods require that the elastic properties and the source location are known. At least six independent measurements (transducers) are necessary for this

NONDESTRUCTIVE TESTING

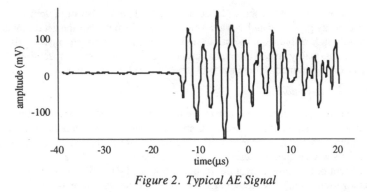

Figure 2. Typical AE Signal

characterization (Scruby et al. 1986). Therefore a multichannel deconvolution routine is necessary to evaluate equation (4). A routine for this is presented below.

Signal Enhancement / Noise Reduction Filters

A problem inherent in any measurement system is the influence of noise on the acquired data. In AE applications the noise can distort the "true" signal, as well as hide low amplitude signals. Noise in this case has two basic sources: electrical noise in the measurement system, and mechanical noise due to material inhomogenieties and other sources which cause the real system to deviate from the ideal assumed in equation (3). This paper will deal only with noise generated by the measurement system.

A typical measured voltage transient from and AE event is shown in Figure 2. In quantitative AE work, the first few cycles of the signal (first few microseconds) are often the only part used for analysis. This is because the transducer resonances typically encountered become difficult to remove beyond this time frame.

An extremely important feature of the AE signal is its time of arrival. The differences in arrival times at an array of transducers is used to establish the location of the AE source. AE source locations have intrinsic interest as they may be used to evaluate damage localization and crack growth. In addition, the source location is necessary to evaluate the Green's function term of equation (2). Since successful quantitative AE analysis hinges upon accurate source locations, the arrival times need to be as accurate as possible. To attain this accuracy the effect of the noise had to be minimized, and the leading edge of the signal had to be magnified.

Noise Reduction:

An analysis of the measurement system noise showed that the noise was essentially a random process. This was determined by calculating the autocorrelation function of a data record containing no signal, but only noise. An ideal random process has an autocorrelation which approaches an impulse function (Chatfield 1989). This autocorrelation was found to be sufficiently close to a delta function to conclude that the system noise is an essentially random (white noise) process (Landis et al. 1992).

The problem with a white system noise is that the frequency components are distributed equally throughout the spectrum so that simple frequency-based filtering is not appropriate. A method was required which is able to reduce the effects of noise, while preserving the features of the actual AE signal. An adaptive moving average filter was found to be effective in fulfilling this requirement.

A simple moving average filter for discrete-time signals may be written:

$$y(n) = \frac{1}{2T+1} \sum_{k=-T}^{T} x(n+k) \tag{5}$$

In this approach each point, $x(n)$, in the signal is replaced by the average of the $2T+1$ data points in the immediate vicinity. An inherent problem with this type of filter is how to set the size of the averaging window. A small window has a little effect on reducing noise, while a large window smears the signal too much. In an adaptive moving average, the window length is adjusted according to the properties of the signal in that region. In this case the window length was set according to the following:

$$T(n) = \frac{1}{a+b\sigma^2} \tag{6}$$

where σ^2 is the sample variance of data points in the vicinity n, and a and b are constants. The constants were determined by setting a maximum T of 16 data points, (1 μs), for a variance of zero, and a minimum T of 3 points, (0.1875 μs) for a variance of two or greater.

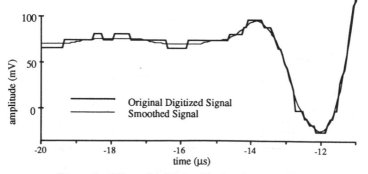

Figure 3. Effect of Adaptive Moving Average Filter

In order to fully optimize this approach, an additional filtering step was inserted prior to averaging. Rabiner and Schafer (1978), showed that a median filter combined with some other smoothing technique can be an optimal method for

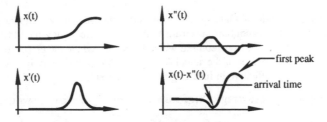

Figure 4. Action of Laplacian Filter

smoothing signals with noise. A running median filter is similar in practice to a moving average, however, the output of a running median filter is the median of the T samples around the data point. Median filters have the ability to remove spikes, while preserving edges. In this application a 3-point running median was used to remove the spikes resulting from digitization. The effect of the combined running median/adaptive moving average is shown in Figure 3.

Signal Amplification:

The second step in locating a precise arrival time is to enhance the signal so that the possible times of arrival can be reduced to a one or two unambiguous points. An effective technique for this purpose is the Laplacian filter. This technique has had successful applications in 2D image analysis for edge detection (Stang et al. 1990). This filter can be written as:

$$y(n) = x(n) - \nabla^2 x(n) \tag{7}$$

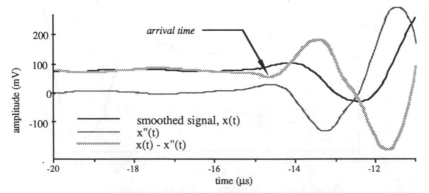

Figure 5. Laplacian Filter and Arrival Time

The effect of this filter is illustrated in Figure 4. As can be seen in the figure, the leading edge of the signal becomes much sharper and the amplitude becomes larger than the original signal. In addition, there is a characteristic "dip" which occurs before the leading edge. The algorithm was coded so that the arrival time was defined as the minimum (or maximum) in the "dip" prior to the leading edge. The effect of this operation on real data is illustrated in Figure 5. It should be noted that implementation of this filter requires a smooth signal to reduce errors occurring in the derivatives of discrete-time signals. The smoothing operations described in the previous section were sufficient for this requirement.

Deconvolution of AE Signals

The inverse procedure of recovering microcrack parameters from surface displacement transients, and the conversion of transducer voltage measurements to displacements require the deconvolution of equation (2). The popular method of frequency division has been shown to be very difficult for AE signals, especially in the presence of noise (Simmons 1985). Time domain techniques, particularly nonlinear least-squares approximations have been shown to be effective for inversion of AE waveforms (Michaels and Pao 1985). This will be the only method discussed here.

The convolution of equation (2) can be written in the discrete-time case as:

$$u^q(n) = \sum_{k=1}^{n} H_i^q(n-k)c_i s(k) \tag{8}$$

where u^q is the normal surface displacement at transducer q, H_i^q is the ith component of the Green's function for the source-receiver pair of transducer q. The temporal and spatial components of the AE source tensor, D_i, in equation (2) have been separated such that $D_i(t) = c_i s(t)$. The goal here is to evaluate the components of c_i and the sequence $s(k)$ for a known set of displacement sequences and Green's functions.

The least squares formulation of equation (8) is:

$$E^q = \sum_{n=1}^{N} \left\{ u^q(n) - \sum_{k=1}^{n} c_i H_i^q(n-k)s(k) \right\}^2 \tag{9}$$

where E^q is the squared error for displacement sequence (channel) q, and N is the number of points in each time series. The minimization criterion is to find c_j and $s(k)$ such that:

$$\sum_{q=1}^{NC} \left\{ 0 - E^q \right\}^2 \tag{10}$$

is a minimum. Here NC is the number of channels used. In order that equation (10) is a minimum, the following must be satisfied:

$$\frac{\partial}{\partial c_i} \sum_{q=1}^{NC} \left\{ 0 - E^q \right\}^2 = 0 \qquad (11)$$

and

$$\frac{\partial}{\partial s(k)} \sum_{q=1}^{NC} \left\{ 0 - E^q \right\}^2 = 0 \qquad (12)$$

The Levenberg-Marquardt algorithm for nonlinear least squares problems was used to evaluate equation (10). A routine published by Press et al. (1986), was found to be very effective, and usually converged in less than ten iterations.

It should be noted that a unique solution is not obtained from this procedure. Some assumptions about the source time function were made to assure uniqueness. Since $s(k)$ is a non dimensional scalar function, it was assumed that the maximum value was $s(k)=1$. This assumption was made so that the resulting source tensor components had the correct absolute magnitude.

The entire analysis routine is summarized in Figure 6.

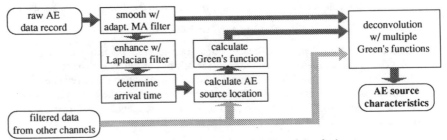

Figure 6. Flowchart for AE Signal Analysis

Applications

The routines described in the previous sections were developed in response to needs which arose out of an experimental program of concrete fracture testing. As a part of this investigation quantitative acoustic emission techniques were applied to investigations of microcracking and damage localization in cement-based materials. The goals of the program were to characterize microcracks in materials of varying composition, and to establish a relationship between microcrack characteristics and bulk material properties such as fracture toughness. Some of the results which were made possible by the analysis routines described in the previous sections are shown below.

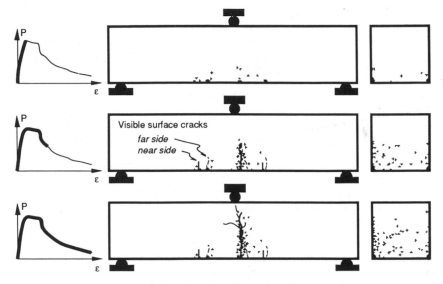

Figure 7. Source locations for mortar beam specimen.

A mortar beam was loaded in three point bending under closed-loop control. The strain (measured by a 4" gage length extensometer) on the tension face of the beam was used as the feedback signal. The dimensions of the beam were 44 x 10 x 10 cm. The span was 40 cm. Eight acoustic emission transducers were mounted on the beam. The AE signals were recorded by a LeCroy modular data acquisition system (LeCroy model TR8837F modular transient recorders housed in a model 8013A CAMAC mainframe). The digitizing rate was 16 MHz at 8 bit resolution. The number of points recorded in each channel was 1024.

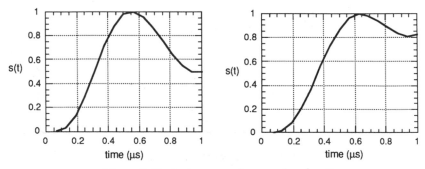

Figure 8. Typical recovered source-time functions

Using the filtering routines described above, for each event the arrival times at each channel were determined, and the source location was calculated using equation (1). These source locations are shown in Figure 7. Localization of damage can be observed to some extent in this figure. In the prepeak region (as highlighted in the load-strain curve on the left side of the figure) the events are located primarily at three separate regions. For increasing strains, however, the additional events are confined to the central crack region. The two crack bands at either side of the central crack were likely initiated by very small shrinkage cracks. As expected, the central crack eventually became the critical crack. The events shown in the cross sectional view illustrate how damage tends to accumulate at the specimen edges first, then moves in towards the center.

Figure 8 shows two typical AE source-time function recovered from the quantitative AE analysis. The source-time function describes the rate at which a microcrack opens (or closes) during an AE event. As can be seen in this figure, the microcracks tend to open at a rate of about 0.4 to about 0.7 microseconds.

Recovered microcrack orientations are shown in Figure 9. All of the microcracks shown are inclined with at least one of the three specimen axes. Even though the overall fracture of the specimen was mode I (tension), the microcracks were found to be primarily mixed mode and mode II (shear).

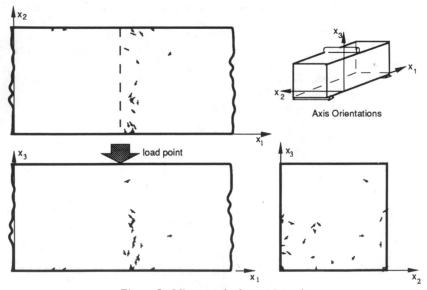

Figure 9. Microcrack plane orientations.

Summary

A brief review of some signal processing techniques with quantitative acoustic emission applications has been presented. This was not intended to be comprehensive, but merely a summary of some of the techniques which have found to be useful for this particular research program. Some applications of the analysis techniques showed that quantitative AE analysis can be a unique tool for evaluating fracture behavior.

Acknowledgment

The research reported herein was supported by a grant from the U. S. Air Force Office of Scientific Research under a program managed by Dr. Spencer T. Wu. Additional support was provided by the NSF Center for Advanced Cement-Based Materials.

References

Aki, K. and P.G. Richards (1980). *Quantitative Seismology*, W.H. Freeman and Company, New York.

Chatfield, C. (1989). *The Analysis of Time Series*, 4th edition, Chapman and Hall, New York.

Enoki, M. and T. Kishi (1988). "Theory and Analysis of Deformation Moment Tensor Due to Microcracking." *International Journal of Fracture*, 38, 295-310.

Landis, E., C. Ouyang, and S.P. Shah (1992). "Automated Determination of First Arrival Time and AE Source Location." *Journal of Acoustic Emission*, 10, S97-S103.

Landis, E.N. and S.P. Shah (1993). "Recovery of Microcrack Parameters in Mortar Using Quantitative Acoustic Emission." accepted for publication in *Journal of Nondestructive Evaluation*.

Hsu, N.N., J.A. Simmons, and S.C. Hardy (1977). "An Approach to Acoustic Emission Signal Analysis." *Materials Evaluation*, 35, 100-106.

Michaels, J.E., and Y.H. Pao (1985). "The Inverse Source Problem for an Oblique Force on an Elastic Plate." *Journal of the Acoustical Society of America*, 77, 2005-2011.

Ohtsu, M., M.Shigeishi, and H. Iwase (1989). "AE Observation in the Pull-Out Process of Shallow Hook Anchors." *Proc. of JSCE* No. 408/V-11.

Ouyang, C., E. Landis, and S.P. Shah (1991). "Damage Assessment in Concrete Using Quantitative Acoustic Emission." *Journal of Engineering Mechanics*, 117(11), 2681-2698.

Press, W.H., B.P. Flannery, S.A. Teukolsky, and W.T. Vetterling (1986). *Numerical Recipes*, Cambridge University Press, New York.

Rabiner, L.R. and R.W. Schafer (1978). *Digital Processing of Speech Signals*, Prentice-Hall, Inc., Englewood Cliffs, New Jersey.

Scruby, C.B., K.A. Stacy, and G.R. Baldwin (1986). "Defect Characterization in Three Dimensions by Acoustic Emission." *Journal of Applied Physics D: Applied Physics*, 19, 1597-1612.

Simmons, J.A. (1985). "Deconvolution for Acoustic Emission," *Review of Progress in Quantitative Nondestructive Evaluation Volume 5A*, D.O. Thompson and D.E. Chimenti, Eds., Plenum Press, New York, 727-735.

Stang, H., B. Mobasher, and S.P. Shah (1990). "Quantitative Damage Characterization in Polypropylene Fiber Reinforced Concrete." *Cement and Concrete Research*, 20, 540-558.

TENSION TESTS OF ARAMID FRP COMPOSITE BARS
USING ACOUSTIC EMISSION TECHNIQUE

Z. Sami[1], H. L. Chen[2] and H. V. GangaRao[3]

Abstract

Performance evaluation of composite rods used as reinforcing bars in concrete structures has been undertaken using acoustic emission (AE) technique. The AE technique has shown promise as a method to establish response characteristics of Aramid Fiber Reinforced Plastic (AFRP) bars under different axial tension levels.

This paper discusses the experimental gripping mechanisms used to test AFRP specimens for direct tension, the AE source location method and the procedure used to eliminate extraneous noises. The behavior of AFRP bars under different tension levels is also discussed in terms of AE parameters, such as number of events and event duration. Finally, this paper discusses a methodology to predict stress levels of AFRP bars using AE parameters, so that this technique can be applied to monitor concrete structures post-tensioned with FRP tendons.

Introduction

The use of Fiber Reinforced Plastic (FRP) composites in Civil structures is being viewed as a serious option to the use of conventional steel reinforcement.

---

[1]Graduate Research Assistant, Department of Civil Engineering, Constructed Facilities Center, West Virginia University, Morgantown, WV 26506.

[2]Associate Professor, Department of Civil Engineering, Constructed Facilities Center, West Virginia University, Morgantown, WV 26506.

[3]Professor of Department of Civil Engineering, and Director of Constructed Facilities Center, West Virginia University, Morgantown, WV 26506.

The advantages of synthetic FRP materials are: (1) higher fiber resistance against acidic corrosion; (2) easier tailoring of the FRP composites for specific application; and (3) higher strength to weight ratio than conventional materials. Since the use of FRP bars in both reinforced and prestressed concrete structures is a relatively new concept, several unresolved technical issues such as ductility, creep and thermal sensitivity have to be researched before implementing these rods in the field. Moreover, design of suitable grips presents a major challenge either for temporary use in pre-tensioned applications or testing FRP bars in tension, or for permanent use in post-tensioned applications. The challenge is mainly encountered by the possible FRP bar failure in the grips under combined effects of shear and compressive stresses or by small rotation or mis-alignment near the grips leading to bending failure.

Two anchorage methods have been previously tried for prestressing aramid rods placed both in pre-tensioned and post-tensioned concrete beams (Kakihara, et al. 1991). They are designated as the wedge method and the grout method. The wedge method employs a couple of wedges with an outer cone. In the grout method, rods are inserted into resin or cement materials to form a cylindrical cone, and it has been shown that AFRP bars have a higher bond strength to concrete than shown by pre-stressed concrete steels. Tests were conducted on the performance of a resin socketed anchor for synthetic prestressing tendons (Holte, et al. 1993) that show that an anchor with a bond release between the resin plug and the metal socket produces a substantial reduction in the inter-laminar shear stress at the tendon surface. The literature review did not reveal the availability of simple reusable grips that are suitable to test the bars in the laboratory. Several methods for anchoring the FRP bars have been investigated (Faza, 1991) to test glass FRP bars under uniaxial tension, including a wet-sand gripping mechanisms (WVU grips) that allows the gripping force to be redistributed during the initial loading stage. In this study a modified wet-sand gripping method (Epoxy-Sand Coated Grips) was developed in order to hold the AFRP bars up to failure at around 12,500 lbs in tension.

The FRP bars in structural applications can experience several types of damages that also may occur at numerous locations. Determining a state of damage in a composite bar requires an assessment of the nature and the quantity of the damage. Several nondestructive methods have been used to monitor the damage occurrences in composites. Among these methods, Acoustic Emission (AE) has been increasingly favored (Hamstad 1986). Composite rods for construction applications were evaluated using AE technique (Chen, et al. 1992). AFRP bars were found to be the strongest when compared to the glass and carbon FRP bars (Uomoto, et al. 1988). It was also observed that virtually no internal failures initiated until the load has reached about 75% of the failure load under constant load rate (not sustained load).

Various composite materials were studied using AE peak amplitude distribution. The peak amplitude distribution has been recognized as a descriptor of the AE signal (Pollock 1981). The study found that expressing AE ring down counts and energy as functions of load can assist in analyzing crack behavior, loading history, and stiffness of concrete beams. AE wave attenuation has also been evaluated as a function of fiber volume fraction for short glass FRP (Choi, et al. 1990). The results showed that for noncontinuous chopped glass FRP the attenuation effect is considerable, and that the peak amplitudes of the AE waves were found to increase as the specimen deformation proceeded.

AE monitoring of concrete structures is difficult because of the heterogeneous nature of concrete. The rate of occurrence of AE activity, however, has been used to predict the extent of internal damages caused by external loads (Maji, et al. 1988, Uomoto 1987, and Ohtsu 1987). AE has been used to determine crack types and orientations, and has demonstrated great promise for source characterization (Ouyang, et al. 1991, Chen, et al. 1992, Maji, et al. 1988 and Ohtsu 1988). Research performed at the Constructed Facilities Center, West Virginia University has indicated that the AE technique can be used to identify the characteristics of different structural materials (Chen, et al. 1992).

The objective of this research was to develop an experimental technique using Acoustic Emission technique to study the behavior of AFRP bars subjected to direct tension. The work described in this paper includes the development of mechanical grips to hold the #2 AFRP bars in tension and the AE test results. This research may be helpful in future development of stress monitoring of AFRP tendons in the post-tensioned prestressed concrete structures.

Experimentation

AE Measurement System

The experimental setup of the tension test is shown in Figure 1. The AE measurement system, AET 5500 (Hartford Steam Boiler Inc.), has two data acquisition channels. Two piezo-electric transducers (Model AC175L) each with a resonant frequency response of 175 kHz and a sensitivity of -70 dB or better have been used. Each AE sensor was connected to a filter and a differential pre-amplifier (Model 160B) with a 60-dB gain. The second amplification stage (post-amplification gain) was set around 14 dB in the AET mainframe.

Successful AE testing relies on separating the true signals from the background noise. Normally the noise observed in static conditions is about 1/6th the observed peak-to-peak noise signal. To set the proper threshold level, AE signals given out by the mainframe for static no load condition were viewed on a storage oscilloscope. The maximum rms voltage was found to be around 65 mV. For a sine wave (AE signal is not a sine wave but is assumed to be one) Voltage rms is 0.707 V peak to peak; hence, the AE signal viewed on an oscilloscope is multiplied by 10 and set at 650 mV for the threshold voltage.

Test Specimen

Aramid FRP bars of #2 size (0.236 in. in diameter) and 36 in. in length were studied under uniaxial tension using the AE technique. The aramid bars were newly developed rods that were obtained by impregnating a straight bundle of an aramid fiber made of poly-phenylene-oxydiphenylene terepthalamide with vinylester resin using the pultrusion method (Kakihara et al., 1991). These specimens were mainly used in prestressed concrete structures, magnetic levitation transportation systems, etc. Some of the mechanical properties of AFRP rods obtained by Kakihara et al. (1991) are shown in Table 1.

Gripping Mechanism

The design and development of suitable grips for AFRP bars under uniaxial tension were conducted. An ideal grip is easy to handle and can be re-used. The grips must grasp the bar when loaded in a manner that prevents the failures of the bar from occurring near the grips. FRP bars usually fail at the grips due to the combined effects of compressive and bending forces. After extensive investigation (Faza, 1991) of holding methods for the rough surfaced glass FRP rods with diameters 0.394 in. or higher at the jaws of Baldwin testing machine, special grips (WVU Grips) of 7 in. x 3 in. and 0.75 in. in thickness were developed. These grips were made out of steel and are shown in Figure 2. A semi-circular groove is cut out of each grip plate. The groove diameter is 1/8 in. greater than the size of the bar being tested. A top piece was attached to the plates, in order to seat them properly inside the jaws of the Baldwin testing machine. Wet sand was filled in the semi-circular grooves of the grips. The bar was then placed within the groove and between the grip plates. These grips achieved some success in testing glass FRP bars, however, these grips can not successfully hold the AFRP bars up to failure.

During the testing of # 2 AFRP bars, the same kind of grips were employed with wet sand used to fill the grooves. All the specimens were observed to slip at the grips at around 3000 lbs of loading. The slippage could have occurred due to the smaller dimension and surface of the bars, which is somewhat smoother than the surface of the glass FRP bars. Overall, several methods of anchoring AFRP bars were investigated, including:
1. Wet-sand filled grips.
2. Wet sand in the grooves of WVU grips with a cement chunk at the ends of the bars outside the grips.
3. Cement between the AFRP bars and the semi-circular grooves of the grips.
4. A thin layer of sand and epoxy paste coating on the surfaces of the semi-circular grooves and the AFRP bars.

AFRP bars are associated with lower Young's modulus (above 18×10^6 psi) and lower toughness than steel bars. The first three techniques did not function satisfactorily for the following reasons:

1. In the sand grip approach, the bars have low radial stiffness and therefore the sand can not exert much pressure on the surface of the bars (the bars will shrink which reduces the area available to develop adequate frictional resistance.)

2. In the case of a hardened cement block placed at the end of bars, the hardened cement paste failed under the transferred tension force because of high tensile strength of the AFRP bars at failure. As a result, it broke loose and slippage of the bar occurred.

3. Several types of cement pastes such as anchor cement and quick setting cement were applied directly in the grooves between the grips and the AFRP bar. During testing, however, failure caused by the bending of the bars was observed near the grips even at low loading levels. Also, after the cement has hardened, the cement paste shrinks, which limits the amount of radial stress to the bar.

The last kind of grips tried consisted of applying a thin layer (about 0.0394 in. thickness) of sand and epoxy paste mixture to the surfaces of the semi-circular groove. The same paste was also applied on the specimen surfaces, which creates a rough surface. Wet sand was then placed between AFRP bars and semi circular grooves. These special "epoxy-sand coated grips" have successfully held the specimens till failure, i.e. up to 12,500 lbs. It was also shown that failure took place always at the middle region of the AFRP bars. The reason for the success of this grip has been attributed to the fact that the coating can provide high strength. It can increase the frictional resistance of the gripping mechanism as the AFRP bar and groove surfaces were roughened after epoxy-sand coating. The rough surfaces help hold the sand in place and help redistribute the stresses, and thus eliminate the bending problem of the AFRP bar (because of wet-sand adjustment inside the grips.) It is noted that steel wedges, as shown in Figure 1, can provide higher compressive stress in the grips for increased loading.

Sample preparation time has been minimized through the use of epoxy-sand coated grips. Three to four tensile tests can be performed with each of the epoxy-sand coated grips. For additional usages of these grips, the special coating in the grooves must be cleaned by heating the coating, and then the same grips can be given a fresh epoxy-sand coating before reusing them.

Test Procedure

The AFRP rods were loaded on a Baldwin testing machine. The major problem encountered was to hold the specimens at high tension levels without slippage occurring. The epoxy-sand coat gripping mechanism explained in the previous section was employed. No slippage and no failure at the grips were observed visually during the tests. Two AE sensors were mounted 18 in. apart on the cleaned surface of each specimen (as shown in Figure 1). The bonds strength between each sensor and the surface of the specimen were kept as equal as possible. Before the beginning of each test, a calibration test was performed to determine the maximum time of travel (maximum DT) from one sensor to another. This was done by a pencil-lead (0.0197 in. dia., 0.118 in. long, HB) break test outside the region. The maximum DT was found to be 75 - 76 μsec. The load was applied gradually and the AE signals from the specimen were recorded through the AE mainframe.

Acoustic Emission Analysis

A linear location (test mode) method, which locates the AE sources along a straight line between two sensors and automatically eliminates the AE sources which are out of the region, such as grips, etc, was employed during testing. This method operates in a time difference mode (TDM). The TDM compares the time of arrival (the time when the first point of the signal is detected above the threshold) of an AE event at two different sensors. The difference between the time of arrival of a signal at one sensor, and the time of arrival of the same signal at another sensor, is referred to as delta T (DT). From the calibration test, the maximum DT was found at around 75 - 76 μsec. In actual testing, the mainframe automatically calculates the location of the AE signals using the maximum DT, sensor spacing and sample DT by a formula given as:

$$Location = (Max.\ DT - DT)\ /\ 2 \times (Sensor\ Spacing/Max.\ DT) \qquad (1)$$

Due to the attenuation of the AE signals, the arrival time will be delayed. This occurs most often when the signals are coming from a location nearer to one of the sensors, since the signals have to travel a longer distance to reach the other sensor. The AE events happening at the mid-region of the two sensors will attenuate in both directions before reaching the sensors. Hence, the measured AE source locations at the middle zone are more accurate than those locations closer to the AE sensors. For this reason, only the events from the middle 60% of the sensor spacing were considered for our data analysis.

Results and Discussion

Six AFRP bars were tested under uniaxial tension and the results are discussed below. Figure 3 shows the distribution of AE events versus location. It can be seen that, AE signals were generated at many different locations on the bar and the failure of the AFRP bars was distributed along the length of the bar, which is advantageous in minimizing the catastrophic failure. Conventional steel bars usually fail at a specific location after necking. All the #2 AFRP bars tested have a failure strength between 12,000 and 13,000 lbs.

Figure 4 shows the time history of AE events. Each data point represents the total AE events for an interval of 10 seconds. This figure indicates changes in rate of AE events for a typical #2 AFRP specimen. The AE event time history can be broadly divided into two stages: Stage I corresponds to loading from zero to 9,500 lbs (time around 300 sec) and Stage II from 9,500 lbs up to failure. Acoustic emissions from the stressed composite materials may occur due to many reasons, such as breaking of the epoxy coating, unwrapping of the fibers, breaking of the matrix, debonding between matrix and fibers, and the breakage of fibers. The maximum rate of events is seen to occur in Stage I. The large scattering in the event rate with couple of peaks seen in Stage I preclude any qualitative conclusion in Stage I. It can perhaps be inferred that Stage I corresponds to failure from breaking of epoxy coating, unwrapping and debonding between matrix and fibers. As the loading reaches the end of Stage I, event rate decreases and reaches a minimum. As the loading continues (after 300 seconds in this case), again the event rate starts to increase gradually. These events seem to result from the breakage of fibers and can be identified using other AE parameters. Note that 9500 lbs load in our tests corresponds to 70% - 75% of the ultimate load of #2 AFRP rods.

Figure 5 shows the plot of mean event duration versus time (or load) for a typical #2 AFRP specimen. Note that this plot closely corresponds to the plot of event rate. It is seen from this figure that the event duration at early stage of loading is high and fluctuating, which indicates that the AE signals are generated from different kinds of failures of the specimens. As the loading reaches around 8500-9500 lbs (corresponds to the end of Stage I in events rate plot), the mean event duration begins to fall and reaches a minimum value. Upon further increase of loading up to failure, the mean event duration remains almost constant, indicating a different failure mechanism. From this observation, it can be inferred that, #2 AFRP bars have undergone failures in two stages. During the first stage, the failures are due to epoxy breaking, debonding, etc., and the AE signals have higher event durations. Whereas in the second stage when the actual fiber breakages are believed to occur, the AE signals typically have lower event durations. This conclusion closely corresponds to the results from Uomoto (1988) i.e., virtually no internal failure of the AFRP rods was noted until the load reached was within 25% of the fiber breakage level.

Event duration versus peak amplitude are plotted separately for Stage I and Stage II in Figures 6(a) and 6(b). Each data point represents a single AE event. Note that AE event numbers in Stage I are higher than those of the Stage II, i.e., less number of events are given out by AFRP bars at the time of the second stage which may correspond to fiber breakages. It is also seen from these figures that the event duration in Stage I has a wide range of variation, whereas, in Stage II the variation was less. The peak amplitudes in both stages were widely distributed.

This observation indicates that the shape of event duration versus peak amplitude may be helpful in identifying the different failure modes in AFRP bars.

The ringdown counts are related to the damping characteristic of the sensor, coupling efficiency and sensor sensitivity. Maintaining stability of these parameters is essential for consistency of interpretation. In general, ringdown counts provide an indication of source intensity or damage severity. The normalized total ringdown counts were plotted against the normalized applied load for all the test specimens. The shape of the curve is similar in all the test specimens. A change in the slopes of the curves was seen at around 50% of the normalized load (Chen, et al. 1993). The trend of the ringdown counts can be in terms of a bi-linear curve with the first slope of 1.6 and the second slope of 0.4. Therefore a drop of the slope to 25% of the initial slope is found to be an important indication of the loading over 50% of the ultimate strength of the AFRP bars.

Application

Current prestressed concrete structures using steel tendons are susceptible to salt corrosion. Composite tendons, on the other hand, do not face the same problem. Use of composite materials such as the AFRP rods (Kakihara, 1991) may significantly increase structure's service life. Tests were conducted by Kakihara (1991), to evaluate the performance of AFRP bars used as tendons in prestressed concrete beams. The results have shown that AFRP bars are useful in place of prestressing steel in severely corrosive environments. It is worth noting that because of the anti-corrosion advantage of AFRP bars, they also have great potential in cable-stayed and suspension bridges. The AE study in this paper has indicated successful prediction of stress levels in the tensioning of AFRP bars. The AE analysis of experimental data can be applied in predicting the stress levels of AFRP bars used as PC tendons in post-tensioned concrete structures where these PC tendons are not in contact with the concrete. AE technique can also be used to study the performance of such structural elements. Further study is necessary to identify signals from concrete cracking and environmental background noise, which is necessary for the development of a health monitoring technique for prestressed concrete structures using AFRP bars as tendons.

Conclusions

Acoustic emission technique was used to monitor aramid FRP bars under uniaxial tension. Epoxy-sand coated grips having an inner layer of sand and epoxy were developed and found to be ideal to hold #2 AFRP bars to failure. Acoustic emissions from AFRP bars were analyzed in detail for proper source location. Distributed failures along the length of the bars was observed and verified through AE source location. AE event rate showed two stages of failure and was able to detect the beginning of fiber failure (after 70% -75% of the ultimate load). The acoustic emissions from breakage of epoxy coating, debonding between matrix and fibers were found to have considerably high event duration than the acoustic emissions from fiber breakage failure. Event duration versus peak amplitude curves are found to be useful in identifying different failure stages of the stressed AFRP bars. Total ringdown counts rate indicates a useful trend in getting early warning of the failure of the AFRP bars. Results of AE parametric analysis were used to predict the stress levels of aramid FRP bars in tension, which will help in the development of a monitoring system for prestressed concrete structures using post-tensioned AFRP tendons.

Acknowledgements

The authors appreciate the support from the NSF and WVDOH through the Constructed Facilities Center at WVU (NSF Grant No. MSS-9016459 & WVDOH Grant No. CIDDMOC). Appreciation is also extended to Mr. Harry M. Mahn, Teijin America, Inc. for supplying the AFRP test specimens.

References

1.   Chen, H. L., Sami, Z., and GangaRao, H. V., to be published on Nov. 1993, "Identifying Damages in Stressed Aramid FRP Bars Using Acoustic Emission" International Symposium on Dynamic Characterization of Advanced Materials, 1993 ASME Winter Meeting, New Orleans, LA.

2.   Chen, H.L., Cheng, C.T., and Chen, S. E., July 1992, "Determination of Fracture Parameters of Mortar and Concrete Beams by Acoustic Emission", Materials Evaluation, Vol. 50, No. 7, 888-894.

3.   Chen, H.L., Sami, Z., and GangaRao, H. V., April 1992, "Acoustic Emission of FRP Bars and A FRP Reinforced Concrete Beam", Proceedings, Fourth International Symposium on AE from Composite Materials, Seattle, WA, 155-165.

4.   Choi, N., Takahashi, K., Oct. 1990, "Characteristics of Acoustic Emission Wave Attenuation in Short Fiber Reinforced Plastics", Journal of Composite Materials, Vol. 24, 1012-1028.

5.   Faza, S.S., Oct. 1991, "Bending and Bond Behavior and Design of Concrete Beams Reinforced with Fiber Reinforced Plastic Rebars", Ph.D. Dissertation, WVU.

6.   Hamstad, M.A., March 1986, "A Review: Acoustic Emission, a Tool for

Composite-Material Studies", Experimental Mechanics, 7-13.

7. Holte, L. E., Dolan, C. W., and Schmidt, R. J., yet to be published, "Anchoring Synthetic Prestressing Tendons", University of Wyoming, Laramie, Wyoming.

8. Kakihara, R., Kamiyoshi, M., 1991, "A New Aramid Rod for the Reinforcement of Prestressed Concrete Structures", Journal of Composite Mater. 132-142.

9. Maji, A., Shah, S. P., March 1988, "Process Zone and Acoustic-Emission Measurements in Concrete", Experimental Mechanics, 773-782.

10. Ouyang, C., Landis, E., Shah, S.P., Nov. 1991, "Damage Assessment In Concrete Using Quantitative Acoustic Emission", Journal of Engineering Mechanics, Vol. 117, 2681-2697.

11. Ohtsu, M., 1987, "Acoustic Emission Characteristics in Concrete and Diagnostic Application", Journal of Acoustic Emission, Vol 6, No.2, 99-108.

12. Ohtsu, M., 1988, "Source Inversion of Acoustic Emission Waveform", Journal of Struc. Eng./Earthquake Eng., Japanese Society of Civil Engineering, Vol. 5, No. 2, 71-79.

13. Pollock, A.A., 1981, Vol. 7, "Acoustic Emission Amplitude Distributions", International Advances in NDT, 215-239.

14. Uomoto, T., 1987, "Application of Acoustic Emission to the Field of Concrete Engineering", Journal of Acoustic Emission, Vol. 6, No.3, 137-144.

15. Uomoto, T., Nishimura, T., Kato, H., Oct. 1988, "Evaluation of Composite Rods For Construction Applications", Translation of paper presented to Japan's Civil Engg. Academic Society,.

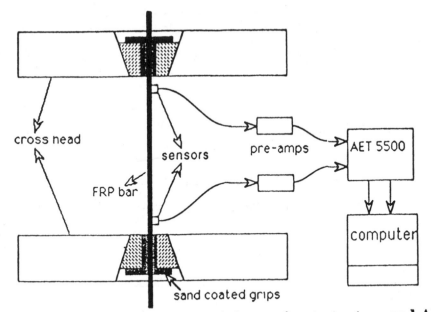

Figure 1 **Schematic diagram of the tension test set-up and AE monitoring system**

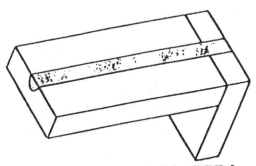

Figure 2 **Sand coated grip to hold #2 AFRP bars while testing**

Matrix Resin	Vinylester
Volume of Fiber	65 %
Density	1.3 g/cubic cm
Tensile Strength	1.9 Gpa
Tensile Modulus	54 Gpa
Elongation	3.7 %
Relaxation	7 - 14 %

Table 1 Mechanical Properties of #2 AFRP Bars

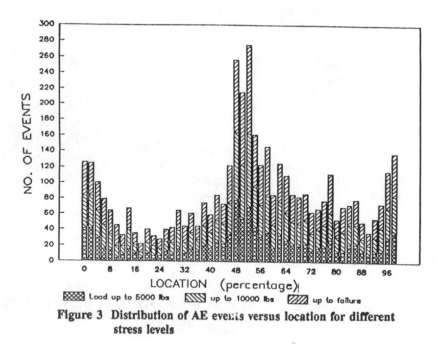

Figure 3 Distribution of AE events versus location for different stress levels

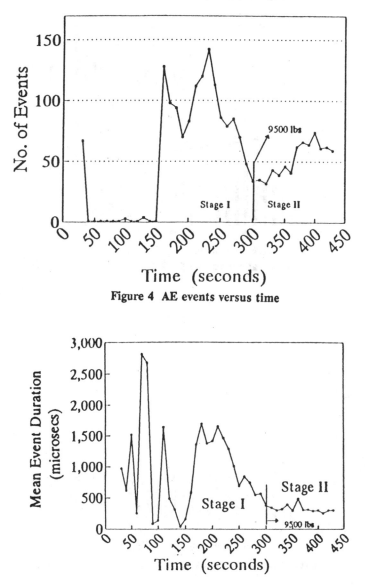

Figure 4 AE events versus time

Figure 5 Mean event duration versus time

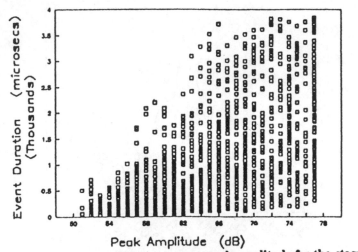

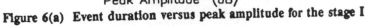

Figure 6(a) Event duration versus peak amplitude for the stage I

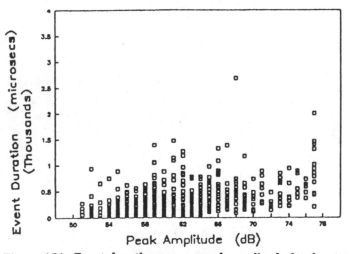

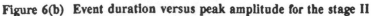

Figure 6(b) Event duration versus peak amplitude for the stage II

CONCEPTUAL DESIGN OF A MONITORING SYSTEM
FOR MAGLEV GUIDEWAYS

Udaya B. Halabe[1], Roger H. L. Chen[1], Powsiri Klinkhachorn[2],
Vasudev Bhandarkar[3], Shen-En Chen[3], Alan Klink[3].

Abstract

The state-of-the-art in nondestructive testing (NDT) technology has
advanced significantly over the past fifteen years and has reached a stage where
continuous monitoring of structures can be successfully achieved. The need for the
adoption of advanced NDT technology for Maglev is absolutely essential because
of the Maglev system safety and performance considerations and high cost of
construction. In this paper, several monitoring methods such as Acoustic
Emission, Ultrasonics, Ground Penetrating Radar, Infrared Thermography, and
Fiber Optic Sensors have been evaluated in terms of their possible use in a Maglev
system. Also, monitoring of guideway alignment and detection of debris on the
guideway using Lasers and Fiber Optics have been investigated.

1. General Monitoring Concept

The two major objectives of monitoring Maglev guideways are early
detection of damages to the guideway system and prevention of catastrophic
structural failures. These objectives can further be integrated into two different
stages of the life of a guideway, namely, during construction and post-construction
periods. Monitoring methods can either be real-time (i.e., continuous with
embedded sensors) or periodic (i.e., intermittent surveying).

[1]Assistant Professor, Department of Civil and Environmental Engineering
[2]Associate Professor, Department of Electrical and Computer Engineering
[3]Graduate Research Assistant
Constructed Facilities Center, College Of Engineering, West Virginia University,
Morgantown, WV 26506-6101.

The essential steps in the monitoring process are identification of defects or damage, determination of suitable methods to predict distress, and determination of steps to minimize, prevent, or eliminate impending distresses.

2. Guideway Defect and Damage Identification

During construction, the monitoring requirements are:

- Curing of the construction material (e.g., maturity and heat of hydration measurements in concrete);
- Time-dependent strength variation (e.g., ultimate growth of concrete strength with time);
- Geometric alignment, with defined tolerance needs; and
- Structural integrity with respect to member joints.

In addition to material aging, the structures are exposed to harsh loading and environmental conditions causing deterioration. Monitoring during the service life of the structure should be focused on:

- Guideway stiffness variations;
- Differential settlements of the foundation;
- Changes in geometric alignment; and
- Integrity of the material.

The main concerns in the applications of NDT to monitor Maglev guideways are the development of NDT techniques which are not significantly affected by the strong electromagnetic fields and the high degree of heat generated during Maglev operations.

3. NDE Techniques Applicable to Maglev Guideways

Nondestructive evaluation (NDE) techniques can be classified broadly into strength estimation, and material properties determination. Strength estimating techniques such as surface hardness methods, penetration resistance methods, etc. are very slow and cause minor localized damages in the structure. Hence, they are not suitable for monitoring Maglev systems. Material property determining techniques, on the other hand, cause no significant disturbances in the structure, can be easily automated, and used as long-term monitoring methods.

4. Acoustic Emission (AE)

AE sensors can be embedded in the structure at the critical locations and can perform continuous monitoring (Ouyang, et al. 1991). However, for concrete guideways, due to the high acoustic wave attenuation and high cost of the sensors, it is advisable to apply metal waveguides to conduct the waves to the sensors. The waveguides are integrated into the structure, particularly over the locations where damages are most critical, and are linked to embedded AE sensors. Hence, a good waveguide would be one that can transmit the signal the farthest and involve the least attenuation. The location of sensors and waveguides are shown in Figure 1.

For FRP (Fiber Reinforced Plastic) and steel reinforced concrete structures, the reinforcement within the concrete structure can be used as waveguides. Waveguides will not, on the other hand, be necessary for a steel structure.

5. Ultrasonics

The levitation magnets are located below the overhang portion of the guideway and support the entire weight of the Maglev vehicle (GangaRao, et al. 1992). This creates tension on the top surface of the guideway near the support of the overhang portion of the guideway, causing the formation of vertical cracks along that line. Vertical cracks are best detected by surface wave transducers, and two sensors are required for each location. Considering six locations, 12 sensors will be required for each section (Figure 2). In addition, positive bending moment at the midspan of the guideway section leads to tension development at the bottom part of the box section. If two surface transducers (not shown in Figure 2) are used to detect any tension cracks at the midspan, a total of 14 sensors per section will be required.

The effect of magnetic fields caused by passing trains on the ultrasonic transducers is temporary, and the interference will fade as the train moves away. In such a case, the output from the sensors will have to be ignored during the time the train is passing over the specific region. However, the long term effects of these high magnetic fields on ultrasonic transducers is not yet known and additional research needs to be undertaken to determine these effects. Corrosion in the case of steel guideway can be detected by determining the thickness of steel members using focussed ultrasonic transducers (Singh and McClintock 1983). However, detection of incipient corrosion and evaluation of corrosion rates by ultrasonics is difficult.

6. Ground Penetrating Radar (GPR)

Electromagnetic radar waves are greatly influenced by the magnetic fields in the Maglev environment, and a separate non-Maglev test vehicle will be necessary for data acquisition. The test vehicle, shown in Figure 3, can run on the guideway at speeds varying from 16 to 32 kmph (10 to 20 mph). Each radar antenna is about 203.2 mm (8 inches) wide and can scan an area of 0.093 m^2 (1 sq. ft.) at a time. Because the distresses will most likely occur along the line of support of the overhang, two antennas will be sufficient to scan the guideway in a single pass.

For GPR, metals such as steel and aluminum have a reflection coefficient of -1 and a transmission coefficient of 0, i.e. radar waves cannot penetrate through metals. Hence, GPR cannot be used as a NDE technique on steel guideways. Also, concrete/composite guideways that are covered with aluminum sheets for shielding of magnetic radiation cannot be monitored by GPR.

7. Infrared Thermography (IR)

The IR system can be mounted on a vehicle for scanning the guideway. However, since the speed of the Maglev vehicle is far too high for the camera to read and record distinguishable temperature differentials, a separate test vehicle which runs at a slower speed [32-48 kmph (20-30 mph)] is necessary for the IR system. The IR system can be mounted on the same test vehicle that houses the GPR system.

Infrared measurements are severely affected by the unpredictable changes in weather conditions (Maser and Roddis 1990, Weil 1991). Defects can be identified only on bright sunny days (or nights following sunny days) during the solar heating cycle (around 11 a.m. - 3 p.m.) or the cooling cycle (around 11 p.m. - 3 a.m.). Furthermore, effect of high temperature gradients due to magnetic fields in the Maglev environment (Campbell and Siu 1988) on infrared measurements has to be properly accounted for in the data analysis.

8. Embedded Fiber Optic Cables to Produce a "Smart Structure"

Conceptually, fiber optic cables can run from inspection stations to the guideway where they are embedded at critical stress areas, and then return to the inspection station. The inspection stations will contain Light Emitting Diodes (LED) that transmit light through the fibers, phototransistors to detect the light intensity passed through the fiber, and support devices (i.e., operational amplifiers, transistors) that amplify the signal for interpretation (Udd 1988). The inspection stations can be positioned either at every joint in the guideway structure, or at intervals of joints as necessary. At least two fibers will need to run the length [about 32 m (105 ft.)] of the guideway. This length does not pose any major problem, since optical fibers with a length of several miles are being successfully used for telephone lines.

Cracks at critical areas will cause deformation of the fiber embedded within the guideway (Figure 4). This deformation in the fiber will allow light which is normally reflected within the fiber to be absorbed in the fiber cladding (microbending loss) (Das, et al. 1985). This will decrease the intensity of light received at the photo-transistor and trigger a warning signal. However, experimental research to determine the minimum crack size that can be detected is yet to be conducted.

The main advantage of using fiber optic sensors is that the strong magnetic field associated with Maglev does not affect the fiber optic cables. The electronic components connected to the fiber optic cables have to be shielded and placed at a safe distance from the magnetic field. As the sensors only need to detect the decrease in light intensity, the data analysis is both easy and inexpensive. Also, strain and temperature variations can be detected by using interferometric devices that measure the phase change in the transmitted light. However, this will require more difficult and expensive data analysis (Jungbluth 1991, Sirkis, et al. 1988).

9. Geometric Alignment using Laser

Monitoring of track geometric variations can provide information on the movement and damage of guideway sections. On a macro-scale, issues such as differential settlements, ground heavings, and horizontal land movements will all result in guideway geometric alignment deviation. On a micro-scale, interior cracks or expansions of a guideway section may result in surface irregularities. These deformations can sometimes be identified by visual inspection. More accurate methods involve the application of a high-tech laser tracking method either locally (mounted on a test car) or globally (placed on the guideway or at a certain distance away from it).

Local laser tracking for geometric alignment has been applied for the purpose of monitoring the alignment of railway systems (CSX Manual 1990). In this application, a test car is equipped with a laser emitting apparatus along with an electronic video camera for receiving the reflected beam. The reflector can be placed permanently on the guideway (Figure 5). The test car runs at a designated speed, typically about 80 kmph (50 mph). An alternative is to install the laser source and camera directly onto the Maglev vehicle. Deformations of the cross section can then be determined by receiving the reflected laser beam by two high-speed cameras. This method is limited to detection of cross sectional misalignment.

In the global system, the laser source and the reflectors are fixed on some extensions on the guideway (Figure 6). By reflecting the light source over the surface of the guideway, adequate information regarding track geometry can be retrieved. This information can be used to detect both longitudinal (along track length) and cross-sectional misalignment. This system has a major advantage in that it does not have to be set up every time.

10. Fiber Optic Guideway Alignment Detector

Fiber optic cables can run from inspection stations to the guideway joints, through deformers, and return to the inspection stations. Any misalignment in the guideway will cause the deformers to press against the optical fibers, inducing microbending losses (Figure 7). This will cause a reduction in the intensity of light received at the phototransistor, and a warning signal will be generated. The amount of light absorbed by the cladding is proportional to the deformation of the fiber.

11. Obstruction Detection on the Guideway

An array of light emitters and receivers (lenses) can be positioned on the guideway (similar to Figure 6). Ordinary light or lasers can be focussed and projected across the guideway by lenses (emitters) and received by a second set of lenses (receivers). The transmitted and received light can be carried from the inspection stations to the lenses and back by fiber optic cables. Any large obstruction (debris) on the guideway will interfere with one or more light beams, resulting in a decrease in the intensity of light at the phototransistor, thus generating a warning signal. The number and positioning of emitters and receivers in the array can be adjusted as necessary.

12. Conclusions

- From the study of various NDT systems, it is evident that a combination of the various NDT methods will be necessary for a complete monitoring system. Such an integrated NDT system combined with advanced data synthesis techniques (e.g., knowledge based systems) can lead to a more accurate and reliable evaluation of maglev guideways.

- The use of an NDT monitoring system can lead to early detection of distresses, thus initiating timely repair steps and preventing the occurrence of major disasters.

- While all the NDT methods discussed are applicable to Concrete structures, Ground Penetrating Radar, Infrared Thermography and Optical Fibers (embedded sensors) are not applicable to Steel structures.

- Research has to be conducted to study the applications of NDT methods to new materials such as Fiber Reinforced Plastics (FRP).

- Research also needs to be conducted to determine the short and long-term effects of strong magnetic fields on NDT sensor/system response.

Acknowledgement

This research was supported by a research grant from the Federal Railroad Administration.

References

1. Campbell, T.I., and Siu, S.W., "Thermal Deformations in Typical Maglev Guideway Structures," Journal of Advanced Transportation, Vol. 21, 1988, pp. 215-226.

2. Das, S., Englefield, C.G., and Goud, P.A., "Power Loss, Modal Noise and Distortion Due to Microbending of Optical Fibers," Applied Optics, Vol. 24, No. 15, August 1985, pp. 2323-2333.

3. GangaRao, H.V.S., Barbero, E.J., Chen, R.H.L., Davalos, J.F., Eck, R.W., Halabe, U.B., Klinkhachorn, P., Raju, P.R., and Spyrakos, C.C., "State-of-the-Art Assessment of Guideway Systems for Maglev Applications", Final Report # DOT/FRA/NMI-92/17, U.S. Department of Transportation, October, 1992, pp. 130-153.

4. Jungbluth, E. D., "Optical Fibers Measure Strain and Temperature", Laser Focus World, January 1991, pp. 155.

5. Maser, K.R., and Roddis, W.M.K., "Principles of Thermography and Radar for Bridge Deck Assessment," Journal of Transportation Engineering, ASCE, Vol. 116, No. 5, Sep./Oct. 1990, pp. 583-601.

6. Ouyang, C.S., Landis, E., and Shah, S.P., "Damage Assessment in Concrete Using Quantitative Acoustic Emission," Journal of Engineering Mechanics Division, ASCE, Vol. 117, No. 11, 1991, pp. 2681-2698.

7. Singh, A., and McClintock, R., "Computer-controlled System for Non-destructive Thickness Measurement of Corroded Steel Structures," Journal of Energy Resources Technology, Vol. 105, December 1983, pp. 499-502.

8. Sirkis, J.S., and Taylor, C.E., "Interferometric-Fiber-Optic Strain Sensor", Experimental Mechanics, June 1988, pp. 170-176.

9. Udd, E., "Embedded Sensors Make Structures "Smart"", Laser Focus/Electro-optics, May 1988, pp. 135-139.

10. Weil, G.J., "Infrared Thermographic Techniques," CRC Handbook on Nondestructive Testing of Concrete, edited by Malhotra, V. M., and Carino, N. J., CRC Press, 1991, pp. 305-316.

11. Track Geometry Manual, CSX Transportation, M/W Services, January, 1990.

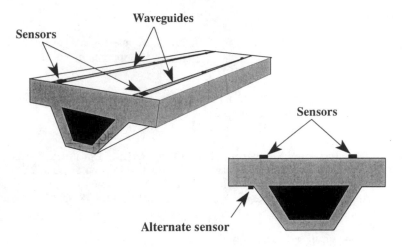

Figure 1.   Acoustic Emission Sensor and Waveguide Locations

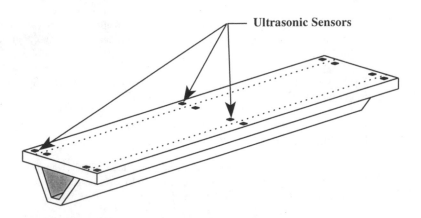

Figure 2.   Positioning of Ultrasonic Sensors on Concrete Guideway

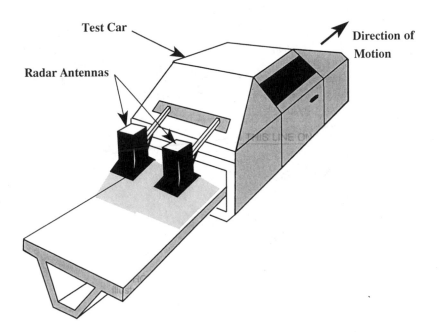

Figure 3. Test Car for Radar Surveying on Guideway

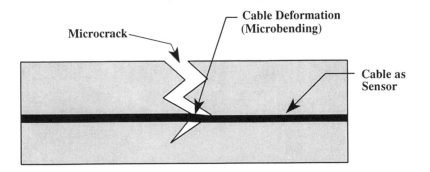

Guideway Section with Magnified Microcrack

Figure 4. Smart Structure Sensor Detecting Microcrack

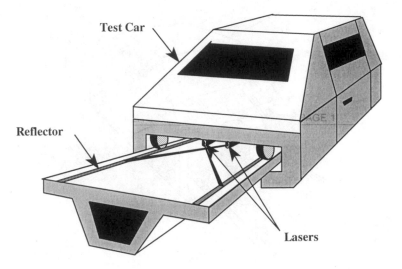

Figure 5. Test Car for Laser Geometric Alignment

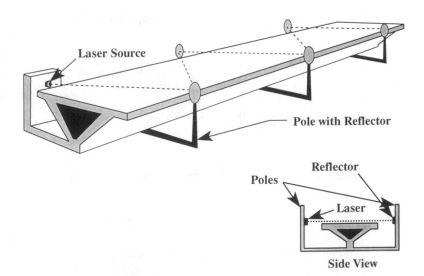

Figure 6. Fixed Laser-Reflector Alignment System
(Also applicable for Debris Detection)

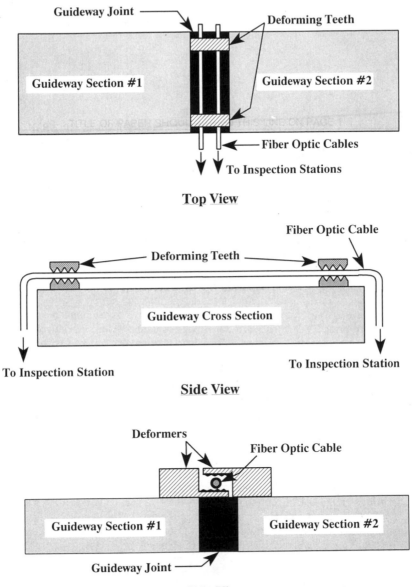

Top View

Side View

Side View

Figure 7.   Fiber Optic Guideway Alignment Detector

NONDESTRUCTIVE TESTING OF A TWO GIRDER STEEL BRIDGE

R.L. Idriss[1], C.B. Woodward[1], John Minor[2] and
D.V. Jauregui[3]

Abstract

Due to geometry and traffic safety considerations, the I-40 bridges over the Rio Grande were scheduled to be demolished in the fall of 1993. This created a unique opportunity for testing. The bridges are classified as "fracture critical", two girder steel bridges. The bridges were tested to determine the effect of defects on the redistribution of loads, on the load capacity, and on the potential for collapse. Two nondestructive techniques, resonant ultrasound and modal analysis, were tested on the bridge. Base line data was accumulated for the undamaged bridge. Then four different levels of damage were introduced in the middle span of one of the girders. Data was accumulated at every stage. The final stage was a near full depth crack in the girder. Static tests and forced vibration tests were performed on the structure.

This is an on-going project. This paper describes the preliminary stages of the project. Full results of the testing, i.e., data obtained from the testing as well as evaluation and assessment of the results will be published at a later date.

Introduction

The I-40 bridges over the Rio Grande in Albuquerque were due to be demolished in the fall of 1993 due to geometry and traffic safety considerations. The bridges represents a common design in the U.S. which is classified as non-redundant, "fracture critical", two-girder steel bridges. The AASHTO Standard Specifications for Highway Bridges[1] defines non-redundant load path structures as structure types "where failure of a single element could cause collapse".

---

[1]Assistant Professor

[2]Professor

[3]Graduate Assistant

These bridges have fatigue sensitive details and are well known to develop fatigue cracks. Experience shows that two-girder highway bridges typically do not collapse when a fracture occurs in a girder. In many instances, they remain serviceable, and damage sometimes is not even suspected until the fracture is discovered incidentally or during inspection[2]. Much still needs to be learned about the after fracture behavior of these structures and how the load gets redistributed when fracture occurs.

Objectives of the Testing

The two main objectives were: a) Load path evaluation of the bridge. Induce a crack and study the load redistribution that occurs in the structure b) Evaluating nondestructive techniques as a tool to detect several levels of damages in the bridge. Two techniques were used: a variation of modal analysis and Resonant Ultrasound, both conducted by Los Alamos National Laboratory.

Bridge Description

The current structure, built in 1963, is a 1,275 foot long, non-composite bridge consisting of three-span continuous units with spans of 131 feet, 163 feet, and 131 feet each. The structural unit is a two-girder welded with bolted splices design with a floor system (Figures 1, 2, and 3).

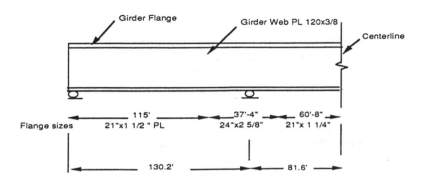

Figure 1 Girder Dimensions

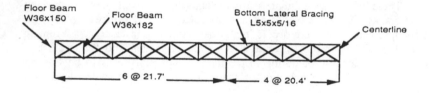

Figure 2 Floor Beams & Bottom Lateral Bracing

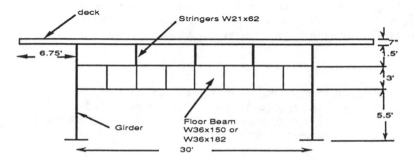

Figure 3 Bridge Cross-Section

Preliminary Analytical Evaluation

A three dimensional finite element computer model of the bridge was developed[3,4]. Defects were modeled in the structure. The results of this analytical study were used to evaluate and choose defects to implant on the bridge. This analysis was necessary from a safety standpoint. The results of the analytical evaluation were used to determine the sensor locations, and optimize the quality and quantity of the data acquisition devices. It was used to calculate the load on the jacks during temporary shoring, the clearance needed at the cut, and the positioning of the truck for static loading. It was also used for the splice design of the girder that was to follow the testing.

Computer Model

A detailed three dimensional analysis of the bridge was necessary to account for the full interaction of the elements and to model the complex three dimensional behavior of the as-built structure (Fig. 4).

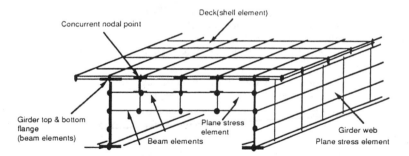

Figure 4 Three Dimensional Model

The Flanges: Beam elements were used to model the top and bottom flanges of the girders, stringers, and floor beams.

The webs were modeled using plane stress elements.

The deck was modeled using shell elements.

Bottom lateral bracing diagonals were modeled as truss elements.

Crack: To model a crack in the girder, coupled nodes were used at all possible crack locations. Except for the fracture location of interest, the other node couples were rigidly linked (Fig. 5).

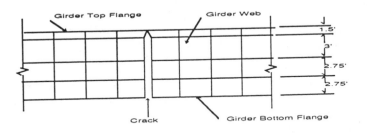

Figure 5 Modeling of the Crack

<u>Systems</u> Two models were examined: (1) a model of the intact structure; (2) a model of the bridge with a crack in one of the girders. A near-full depth crack was modeled at midspan of the central span, extending through the bottom flange and the web. An elastic analysis of both systems was conducted.

To determine the alternate load paths along which the load gets redistributed, the forces and moments generated in the different elements of the structure were evaluated at the vicinity of the crack and at the interior supports.

Results of the Computer Analysis[3,4] When a crack was introduced in the girder there was a shift in load throughout the structure. When the results were interpreted the following were evident:

1.   The introduced crack changed the existing structure into a new, still stable structure.

2.   Most of the load was observed to be redistributed longitudinally, via the girders and the stringers to the interior supports.

3.   Some of the load was redistributed in the transverse direction to the intact girder due to the torsional rigidity of the deck, the floor beam, and the bracing system.

4.   The bottom lateral bracing at the crack location had a large increase in tensile force and played a significant role in stiffening and stabilizing the structure. Bracing at other locations did not show any significant changes.

Monitoring of the Bridge

Based on the preliminary finite element analysis and load path evaluation, we focused on monitoring the elements which showed the most significant change in load. Using strain gages, we primarily monitored the moments at the interior girder supports, forces in the bracing at the vicinity of the crack, moments in the floor beams and stringers at the vicinity of the crack.

Bridge Testing

Static testing: A general tractor trailer was used with a front to back axle spacing of 55.18 ft, and a gross vehicle weight of 81,620 lbs which was 95.5% of the maximum New Mexico legal load. Influence lines for a three span continuous beam were used, and the truck was placed above the north girder at two locations for the static load test: at the center of the middle span to create maximum positive moment in the center span and almost maximum negative moment at the supports, and in the first span to create minimum moment in the middle span.

Dynamic testing: A hydraulic shaker furnished by Sandia National Laboratory, consisting of 21,700 lbs reaction mass supported by three air bearings resting on top of 55 gallon drums filled with sand was used. A 2,200 lb hydraulic actuator located at the center of the mass and anchor bolted to the bridge deck provided the input force to the bridge. A generator powered the pumps necessary to operate the actuator and a water truck provided cooling for the pumps. The shaker could provide either a random or swept sine input. An accelerometer mounted on the reaction mass was used to measure the force input to the bridge. This indirect force measurement gives the total force transferred to the bridge through the actuators and the 55 gallon drums. Forced vibration tests were conducted by Los Alamos National laboratory using a random input, therefore, experimental modal analysis and resonant ultrasound could be performed on the bridge. Piezoelectric accelerometers were used for the vibration measurements. Experimental modal analyses were repeated after each level of damage was introduced in the bridge.

Damage Description

Four different levels of damage were introduced in the middle span of the north plate girder by making various torch cuts in the web and the flange of the girder. This occurred from September 3 through 8, 1993. The final cut was to simulate a near-full depth crack in the girder. This type of crack, usually a fatigue crack, develops at fatigue sensitive details in the bridge, often at the girder to floor beam connection. The cut in the girder was done in four stages.

- The first stage cut was a two feet deep cut in the web originating at the floor beam connection level.

- Next, the cut was continued to the bottom of the web. During this cut, the web bent out of plane approximately 1 inch.

- The third stage was to cut the flange half way in from each side, directly below the web cut.

- Finally, in the fourth stage, the flange was severed completely, leaving the upper four feet of the web and the top flange to carry load at that location. The final cut resulted in a six feet deep crack in the ten feet deep girder, extending from the bottom flange to the floor beam to girder connection.

Preliminary Results

The fractured bridge was stable, as predicted by the preliminary finite element analysis. Under dead load, the bottom flange of the girder at the cut deflected by only 11/16 of an inch and the crack opened 3/8 of an inch. When the truck was positioned above the cut, the crack opened to 3/4 of an inch and the girder at that location deflected an additional 1/2 inch for a total of 1-3/16 inch.

Future Work

The experimental data are currently being analyzed. Data from the testing, as well as evaluation and assessment of this data will appear in a final report that is currently being assembled.

REFERENCES

1. American Association of State Highway and Transportation Officials, "Standard Specifications for Highway Bridges," AASHTO Code, 13th ed., 1983.

2. Daniels, J.H., Kim, W., and Wilson, J.L., "Recommended Guidelines for Redundancy Design and Rating of Two Girder Steel Bridges," NCHRP Report 319, 1989.

3. Idriss, R.L., White, K.R., Woodward, C.B., Minor, J., and Jauregui, D.V., "Bracing, A Secondary Load Path in a Fracture Critical Bridge," Proceedings of the Structure Stability Research Council, "Is Your Strucuture Suitably Braced?," Milwaukee, WI, April, 1993.

4. Idriss, R.L., White, K.R., and Jauregui, D.V., "After Fracture Response of a Two-Girder Steel Bridge," Structure's Congress 1993, Irvine, CA, April 1993.

AN INFORMATION SYSTEM ON THE PERFORMANCE OF SUSPENSION BRIDGES

UNDER WIND LOADS: 1701-1993

Satinder P. S. Puri,[1] M. ASCE

ABSTRACT

From 1701 to 1993, most suspension bridges performed satisfactorily under wind loads. Among those that did not, a few were destroyed, some partially damaged, and others experienced unacceptable vibrations. Suspension bridges with unacceptable vibrations were fitted with various stiffening devices to improve performance. Twentieth Century suspension bridges performed better than 19th Century bridges - since fewer bridges were destroyed or damaged in the present century. However, performance data, where available, varies from brief descriptions to detailed explanations.

Data on performance is being collected for a Computer-Aided Information System, currently under development. The performance of a suspension bridge, under wind loads, is recorded by listing the vibrational response characteristics and the damage, if any. The performance is evaluated for the bridge as a "whole" and for the individual "parts" during the "construction" and "service" stages. While this paper evaluates the performance of full-scale bridges only, the Computer-Aided Information System will include the performance of scaled laboratory models, and models in wind tunnels.

This paper is a work-in-progress report on the performance of 26 full-scale suspension bridges. Five of these bridges were destroyed by wind, and six were partially damaged. Seven bridges that were not damaged had stiffening devices added to improve performance. One bridge was damaged after the devices were added. Another bridge, with a record of satisfactory performance in the unstiffened single-deck state, had bottom-chord lateral bracing added to improve its performance in the stiffened double-deck state. Six bridges experienced, according to their designers, acceptable vibrations.

HISTORICAL BACKGROUND

The first recorded failure of a suspension bridge by a gale was the Dryburgh Abbey Bridge, in Scotland in 1818. This was followed by wind-related failures of the Wheeling Bridge in 1854, the Lewiston-Queenston Bridge in 1864, the Niagara-Clifton Bridge in 1889, and the dramatic failure of the first Tacoma Narrows Bridge in 1940. Despite these failures, spread over a 120-year period, engineers continued to design suspension bridges for the effects of static loads only because the Navier, Rankine, Elastic, and Deflection theories for design of suspension bridges did not provide guidelines for aerodynamic design. While designers were aware of aerodynamic effects, their designs to counter the destructive effects of wind continued to be based on "feel" rather than on "facts."

In the early bridges, railings - either framed or trussed were used to stiffen the deck against vertical displacements caused by moving loads. Fig. 1 shows an unstiffened suspension bridge. Diagonal stays, in various configurations, were used as stiffening devices against wind loads even

----
[1] Structural Engineer, Port Authority of New York & New Jersey, One World Trade Center, Floor-72 N1, New York, NY, 10048

though their behavior was not clearly understood. The first known use of stays was for the Winch Bridge in England, a two ft wide pedestrian bridge with a single span of 70 ft, which opened in 1741. As spans became longer, trussed railings led to the development of stiffening trusses which were used extensively in suspension bridges opened to traffic between 1869 and 1930. Some of them used diagonal stays in combination with trusses. The Clifton Bridge, which opened in England in 1864, was one of the few 19th Century bridges that used plate girders.

Between 1931 and 1947, there was a trend toward graceful looking bridges. As a result, the stiffening truss was either eliminated as with the first stage of the George Washington Bridge or replaced by rolled beams in the Fykesund Bridge in Norway, and by slender plate girders in at least six North American bridges, including the first Tacoma Narrows Bridge.

Twentieth Century bridges with rolled beams or plate girders performed poorly under wind loads. Some 20th Century bridges with stiffening trusses also performed unsatisfactorily. To improve performance, bridges were fitted with a variety of stiffening devices such as: cable ties, diagonal stays, fairings, guy ropes, lateral braces, transverse braces, and dampers. Fig. 2 shows a variety of stiffening devices. While stiffening trusses are still popular, streamlined boxes have been used in some European and Turkish bridges - generally in combination with inclined suspenders.

PERFORMANCE

Performance: a measure of functional effectiveness, refers to performance of a bridge as a "whole" and to performance of the individual "parts" such as deck framing, railings, stiffening devices, suspension elements (chains/wires/strands/cables), suspenders (also known as hangers), and towers. The performance is recorded both during the "construction" and "service" stages of a bridge. The various parts are shown in Fig.'s 1 & 2.

Suspension bridges vibrate in a variety of patterns and modes when subjected to aerodynamic phenomenon such as buffeting, flutter, galloping, and vortex excitation. A suspension bridge can be considered to perform adequately if the stresses, deflections, amplitudes of vibration, and frequency of occurrence of such vibrations are within acceptable limits.

Initially, laboratory tests were used to predict performance. Later wind tunnel tests were used for the same purpose. Nowadays, wind tunnel tests, computer analyses, and simulations are used to predict and to verify performance.

Performance is recorded in the Computer-Aided Information System by listing the vibrational response characteristics of a suspension bridge and the damage, if any, when subjected to wind loads. Vibrational response characteristics are recorded by listing - measurements of frequency, amplitude, mode shape, symmetry of mode shape with respect to the longitudinal and cross-sectional center-lines of a bridge, and the type of vibration: vertical, torsional, or coupled. Fig. 3 shows types of vibrations. Damage is recorded by describing its extent.

While this paper evaluates the performance of full-scale bridges only, the Computer-Aided Information System will include the performance of scaled laboratory models, and models in wind tunnels.

Accounts of performances of earlier bridges are not well documented. Since these accounts, where available, are usually brief and often incomplete, it is difficult to assess how earlier bridges actually performed in the wind environment. However, in a few cases, eyewitness reports, supplemented with sketches, are available. A movie camera was used to record the performance of the first Tacoma Narrows Bridge. Automated data-acquisition devices are often used to record the performances of present-day bridges.

COMPUTER-AIDED INFORMATION SYSTEM

The Computer-Aided Information System, now under development, will contain data on the performance of suspension bridges under wind loads. Complete information has not yet been gathered on all the bridges. In some cases, the information is not available since it was never collected by owners and operators. In other cases, the information is pending because references containing this information are not easily accessible.

Since images take up a lot of storage space, at present, only textual information is being collected. The textual information, extracted from available references, is being organized in a series of tables. For a particular suspension bridge, the system will contain information on items such as general bridge data, performance of full-scale bridges during the "construction" and "service" stages, performance of laboratory models and models in wind tunnel tests, instruments for measuring wind data and response of full-scale bridges, suggestions for improving performance, and a list of references.

Each bridge has a distinct ID consisting of the letter "S" followed by the year in which the bridge was opened to traffic. Letters "A", "B", "C", etc. are used as suffixes to distinguish bridges, in order of decreasing main-span, which opened to traffic in the same year. For example, the ID's: S-1939A and S-1939B denote the Bronx-Whitestone Bridge, NY, USA (main-span = 2300 ft) and the Deer Isle Bridge, Maine, USA (main-span = 1080 ft) respectively - both opened to traffic in the year 1939.

STIFFENING ELEMENTS & DEVICES

Besides traditional elements such as railings, trusses and girders, several devices such as cable ties, dampers, diagonal stays, fairings, guy ropes, lateral braces, stays, and transverse braces, were used to stiffen suspension bridges against the destructive effects of wind. Fig. 2 shows various elements and devices. However, ties, fairings, and braces are not shown in the figure. The elements and devices are components of stiffening systems.

Various references have used terms such as "diagonal stays", "guy ropes", and "stays" interchangeably. Where possible, the information system uses the term "diagonal stay." Diagonal stays have been classified as: anchor stays - stays from concrete anchors in the ground to underside of stiffening elements; corner stays - stays from top of stiffening elements at corners of main or side spans, at the towers, up to the cable bands; mid-span stays - stays from cable bands at mid-span down to top of stiffening elements; pier stays - stays from base of towers or piers up to underside of stiffening elements, and tower stays - stays from top of towers down to top of stiffening elements.

Cable ties connect the main cables and the top of the stiffening elements, in a crisscross pattern, in a vertical plane. Fairings are attachments, usually connected to the deck edges, on either side, to streamline the flow. Lateral braces connect the bottom chords or flanges of stiffening elements in a horizontal plane. Transverse braces connect main cables with brackets mounted at the top of stiffening elements, in the plane of the cross-section.

PERFORMANCE OF SUSPENSION BRIDGES UNDER WIND LOADS

This paper is a report on the performance of 26 full-scale suspension bridges. Performance under wind loads is recorded and evaluated for the bridge as a "whole" and also for the individual "parts" - both during the "construction" and "service" stages. The data on performance is organized in tables.

TABLE 1

TABLE 1 contains a list of 26 suspension bridges included in the system on which some information on performance under wind loads is available. Information is arranged in two rows. The first row contains information on the IDentification mark; NAME, CROSSING, & LOCATION; and the name of the DESIGNER(S). The second row contains information on TYPE of bridge: PEDestrian, HighWaY, PIPEline, or RailRoad; dimensions of MAIN and SIDE-SPANS (unless noted, side-spans are two in number and equal), SAG, CLEARance, and WIDTH of bridge or distance center/center of suspension elements, in feet; type of SUSPension ELEment - chains, wires, strands, or cables; and type of STIFFENING ELEMENT as originally installed - Railings, Girders or Trusses. Subsequent additions, if any, such as Braces, Dampers, Fairings, Stays and Ties - are included in Table 2.

NOTES FOR TABLE 1

Note 1(S-1817):	The single span of Dryburgh Abbey I was supported by a system of inclined chains connected to the wooden deck at several points.
Note 2(S-1823):	The Brighton Chain Pier consisted of a total of four spans, each 255 ft in length.
Note 3(S-1826):	The side spans of the Menai Straits Bridge are supported on masonry arches.
Note 4(S-1889):	While Niagara-Clifton II (replacement for S-1869: Niagara-Clifton I) was based on the original drawings, the bridge was also widened and strengthened.

TABLE 2

TABLE 2 contains outlines of performances of full-scale bridges. Performance is described in terms of whether the bridge experienced VERtical, TORsional, and/or CouPLed VIBrations, and whether it was also DAMAGED during the "CONSTRUCTION" and "SERVICE" stages. The table also includes information on the ORIGinal STIFFENING ELEMENT(S) and subsequent ADDitions, if any. The additions include devices such as cable ties, diagonal stays, fairings, guy ropes, lateral braces, transverse braces, and dampers.

TABLE 3

TABLE 3 contains information on the AGE of suspension bridges that are in operation, as of 1993, and the LIFE-SPAN of suspension bridges that were destroyed by wind, or were demolished, or replaced.

SUMMARY OF PERFORMANCES

Total Number of Bridges   = 26
No of 19th Century Bridges = 12
No of 20th Century Bridges = 14

Bridges destroyed = 5 (Dryburgh Abbey I, Wheeling I, Lewiston-Queenston I, Niagara-Clifton I, and Tacoma Narrows I)
Bridges damaged = 6 (Brighton Chain Pier, Menai Straits, Montrose I, Nassau, Roche-Bernard, and Niagara-Clifton II)
Bridges fitted with stiffening devices to counter excessive vibrations = 7 (Golden Gate, Fykesund, Thousand Islands(USA), Thousand Islands(Canada), Bronx-Whitestone, Beauharnois Dam, and Forth Road)
Bridges damaged after being fitted with stiffening devices to counter excessive vibrations = 1 (Deer Isle)
Bridges with satisfactory performance = 7 (Union, Clifton, George Washington, Lions Gate, Peace River, Liard River, and Severn)

Bridge with shortest life-span: Tacoma Narrows I (4 months)
Oldest bridge in operation: Union Bridge (173 years as of 1993)

Twentieth Century suspension bridges performed better than 19th Century bridges. Fewer 20th Century bridges were destroyed or damaged in this century.

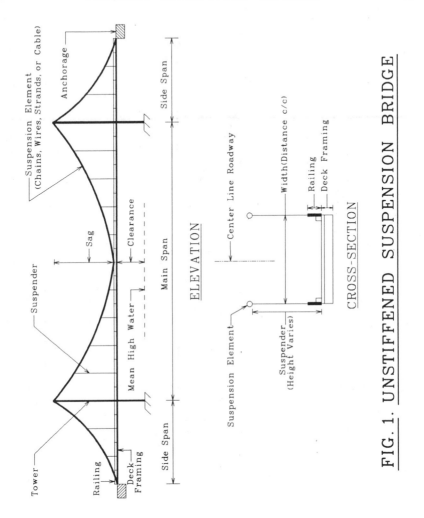

FIG. 1. UNSTIFFENED SUSPENSION BRIDGE

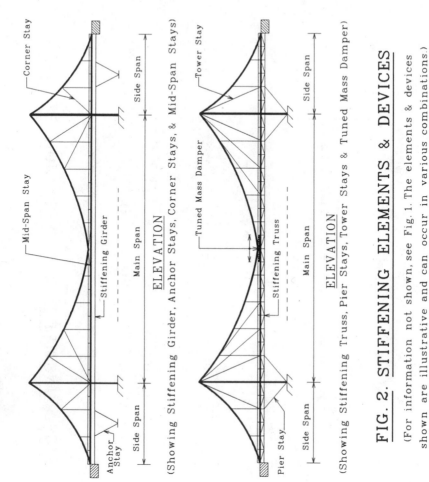

FIG. 2. STIFFENING ELEMENTS & DEVICES

(For information not shown, see Fig. 1. The elements & devices shown are illustrative and can occur in various combinations.)

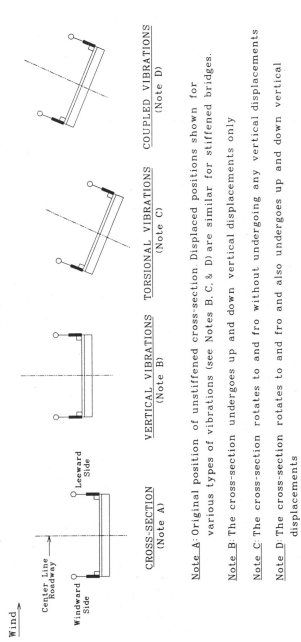

Wind →

Center Line Roadway

Leeward Side

Windward Side

CROSS-SECTION
(Note A)

VERTICAL VIBRATIONS
(Note B)

TORSIONAL VIBRATIONS
(Note C)

COUPLED VIBRATIONS
(Note D)

Note A: Original position of unstiffened cross-section. Displaced positions shown for various types of vibrations (see Notes B, C, & D) are similar for stiffened bridges.

Note B: The cross-section undergoes up and down vertical displacements only

Note C: The cross-section rotates to and fro without undergoing any vertical displacements

Note D: The cross-section rotates to and fro and also undergoes up and down vertical displacements

FIG. 3. TYPES OF VIBRATIONS

(For information not shown, see Fig. 1)

TABLE 1. GENERAL DATA

ID	NAME, CROSSING, & LOCATION									DESIGNER(S)
	TYPE	MAIN SPAN	SIDE SPANS	SAG	CLEAR.	WIDTH	SUSP. ELE.	STIFFENING ELEMENT	W.T. TEST	
S-1817	DRYBURGH ABBEY I, Tweed River, Berwick County, Scotland									Built by J. & W. Smith
	PED.	260	-	?	?	4	*Note 1*	Wooden Railings	-	
S-1820	UNION, Tweed River, Nordham Ford, England/Scotland									Sir Samuel Brown
	HWY.	449	-	30	27	18	Chains	Railings	-	
S-1823 *Note 2*	BRIGHTON CHAIN PIER, English Channel, Sussex County, England									Sir Samuel Brown
	PED.	255	255	18	?	12.67	Chains	W.I. Railings	-	
S-1826	MENAI STRAITS, Wales									Thomas Telford
	HWY.	580	*Note 3*	43.33	100	28	Chains	Railings	-	
S-1829	MONTROSE I, South Esk River, Angus County, Scotland									Sir Samuel Brown
	HWY.	432	-	42	?	26	Chains	Railings - ?	-	
S-1830	NASSAU, Lahn River, Germany									Built by Lossen & Wolf
	HWY.	245.6	38.9	?	?	22	Chains	?	-	
S-1840	ROCHE-BERNARD, Vilaine River, Morbihan, France									Built by Le Blanc
	HWY.	640.6	-	50	110	?	Cables	?	-	
S-1849	WHEELING I, Ohio River, West Virginia, USA									Charles Ellet
	HWY.	1010	-	?	97	24	Cables	Wooden Railings	-	
S-1850	LEWISTON-QUEENSTON I, Niagara River, Lewiston, NY, USA/Queenston, Ontario, Canada									Edward Serrell
	PED.	1043	-	?	?	21	Cables	Railings	-	
S-1864	CLIFTON, Avon River, Clifton, England									I.K. Brunel
	HWY.	702.25	?	?	?	31	Chains	W.I. Plate Girders	-	
S-1869	NIAGARA-CLIFTON I, Niagara Falls, NY, USA									Samuel Keefer
	PED.	1268.33	-	?	?	10	Cables	Wooden Trusses	-	
S-1889 *Note 4*	NIAGARA-CLIFTON II, Niagara Falls, NY, USA									G.W. McNulty
	HWY.	1268.33	-	?	?	?	Cables	Steel Trusses	-	
S-1931	GEORGE WASHINGTON, Hudson River, NY/NJ, USA									O.H. Ammann
	HWY.	3500	610,650	316	213	106	Cables	None	Yes	
S-1937A	GOLDEN GATE, San Francisco Harbor, California, USA									J. B. Strauss
	HWY.	4200	1125	475	220	90	Cables	Steel Trusses	Yes	
S-1937B	FYKESUND, Norway									?
	HWY.	755	-	95.5	?	23.6	Cables	Rolled Steel I-Beams	?	
S-1938A	THOUSAND ISLANDS, St. Lawrence River-American Channel, NY, USA									Hilton Robinson & D.B. Steinman
	HWY.	800	350	?	150	30.5	Cables	Steel Plate Girders	Yes	
S-1938B	THOUSAND ISLANDS, St. Lawrence River-Canadian Channel, Canada									Hilton Robinson & D.B. Steinman
	HWY.	750	300	?	120	30.5	Cables	Steel Plate Girders	No	
S-1938C	LIONS GATE, Vancouver Harbor, British Columbia, Canada									?
	HWY.	1550	614	150	200	40	Cables	Steel Trusses	Yes	
S-1939A	BRONX-WHITESTONE, East River, Bronx/Queens, NY, USA									O.H. Ammann
	HWY.	2300	735	200	150	74	Cables	Steel Plate Girders	Yes	
S-1939B	DEER ISLE, Maine, USA									D.B. Steinman
	HWY.	1080	484	108	85	23.5	Cables	Steel Plate Girders	Yes	
S-1940	TACOMA NARROWS I, Puget Sound, Washington, USA									L.S. Moisseiff
	HWY.	2800	1100	232	203	39	Cables	Steel Plate Girders	Yes	
S-1943A	PEACE RIVER, Alaska Highway, British Columbia, Canada									Raymond Archibald
	HWY.	930	465	93	?	30	Strands	Steel Trusses	No	
S-1943B	LIARD RIVER, Alaska Highway, British Columbia, Canada									Raymond Archibald
	HWY.	542.8	232.7	54.28	?	30	Strands	Steel Trusses	No	
S-1947	BEAUHARNOIS DAM, St. Lawrence Seaway, Quebec, Canada									?
	HWY.	580	-	58	?	30	Cables	Steel Plate Girders	No	
S-1964	FORTH ROAD, Firth of Forth, Scotland									Sir Gilbert Roberts
	HWY.	3300	1340	300	170	78	Cables	Steel Trusses	Yes	
S-1966	SEVERN, England/Wales									Sir Gilbert Roberts
	HWY.	3240	1000	270	120	75	Cables	Steel Box Girder	Yes	

(For description of Notes 1, 2, 3, & 4 - see preceding pages)

TABLE 2. OUTLINE OF PERFORMANCES OF FULL-SCALE BRIDGES

ID	NAME	DURING CONSTRUCTION				DURING SERVICE			
		VER. VIB.	TOR. VIB.	CPL. VIB.	DAMAGED	VER. VIB.	TOR. VIB.	CPL. VIB.	DAMAGED
S-1817	DRYBURGH ABBEY I	?	?	?	No	?	?	?	Yes

STIFFENING ELEMENT(S): Wooden railing.
PERF. DURING SERV.: Completely destroyed by a gale on January 15, 1818.

| S-1820 | UNION | ? | ? | ? | No | ? | ? | ? | No |

STIFFENING ELEMENT(S): Railing. Type - ?
PERF. DURING SERV.: A number of references have stated, incorrectly, that the bridge was destroyed in a gale about six months after it opened. However, the bridge did survive all storms and is still in operation.

| S-1823 | BRIGHTON CHAIN PIER | ? | ? | ? | No | Yes | ? | ? | Yes |

STIFFENING ELEMENT(S): Wrought - iron trussed railing
PERF. DURING SERV.: a. The four-span bridge vibrated during high winds. b. The third span was partially damaged by a gale on Oct. 15, 1833. c. The deck and railing of the third span, which vibrated the most, were again damaged by a gale on Nov. 30, 1836. This time the response of the bridge was witnessed and recorded, with sketches, by Lt. Col. Reid of the British Army.

| S-1826 | MENAI STRAITS | ? | Yes | ? | Yes | ? | Yes | Yes | Yes |

STIFFENING ELEMENT(S): Railing. Type - ?
PERF. DURING CONST.: There were several gales that caused a slight damage.
PERF. DURING SERV.: a. Some suspenders and floor beams were fractured by two heavy gales on Feb. 6, 1826. b. Several suspenders were fractured during a heavy gale in Jan. 1836. c. More than one-third of the suspenders and a large portion of the deck were destroyed by a gale on Jan. 6, 1839. In addition, the chains slapped each other damaging some of the connecting bolts. While the bridge was repaired the previous two times, this time it was also strengthened. d. Nearly a 100 years later, the bridge was again damaged by a gale in Jan. 1936. This time the bridge underwent a major rehabilitation.

| S-1829 | MONTROSE I | ? | ? | ? | No | ? | ? | ? | Yes |

STIFFENING ELEMENT(S): Railing - ?
PERF. DURING SERV.: Damaged by a gale on Oct. 11, 1938. Bridge was repaired.

| S-1830 | NASSAU | ? | ? | ? | No | ? | ? | ? | Yes |

STIFFENING ELEMENT(S): ?
PERF. DURING SERV.: Of the 16 eyebar chains, 8 each side, 12 broke during a wind storm in the winter of 1833-34.

| S-1840 | ROCHE-BERNARD | ? | ? | ? | No | ? | ? | ? | Yes |

STIFFENING ELEMENT(S): ORIG. - ? ADD. - Pier stays.
PERF. DURING SERV.: The deck was damaged by a hurricane on Oct. 26, 1852. The deck was repaired and pier stays were added to improve the performance.

| S-1849 | WHEELING I | ? | ? | ? | No | Yes | Yes | ? | Yes |

STIFFENING ELEMENT(S): Railing. Type - ?
PERF. DURING SERV.: On May 17, 1854, a hurricane destroyed the deck and also pulled 10 of the 12 cables out of the anchorages. An eyewitness account of the failure was reported in a local newspaper.

| S-1850 | LEWISTON-QUEENSTON I | ? | ? | ? | No | ? | ? | ? | Yes |

STIFFENING ELEMENT(S): ORIG. - Trussed railings. ADD. - Pier stays added in 1855.
PERF. DURING SERV.: a. Pier stays were added after the bridge was shaken by a storm in 1855. b. The stays were removed during an ice jam in 1864. Since the stays were not replaced on time, the bridge was destroyed by a storm on Feb. 1, 1864.

| S-1864 | CLIFTON | ? | ? | ? | No | ? | ? | ? | No |

STIFFENING ELEMENT(S): 3 ft. deep wrought-iron rivetted plate girders.
PERF. DURING SERV.: Double amplitudes of more than 6 in. were observed.

| S-1869 | NIAGARA-CLIFTON I | ? | ? | ? | No | ? | ? | ? | Yes |

STIFFENING ELEMENT(S): 6.5 ft. deep wooden stiffening trusses.
PERF. DURING SERV.: Destroyed by a hurricane on Jan. 9, 1889. Replaced by S-1889.

| S-1889 | NIAGARA-CLIFTON II | ? | ? | ? | No | ? | ? | ? | No |

STIFFENING ELEMENT(S): Steel stiffening trusses. Depth - ?
PERF. DURING SERV.: The bridge was damaged by a storm during erection.

| S-1931 | GEORGE WASHINGTON | ? | ? | ? | No | Yes | Yes | ? | No |

STIFFENING ELEMENT(S): ORIG. - None. Stability was achieved thru the heavy self weight. ADD. - 29 ft. deep stiffening trusses and bottom chord lateral bracing were added in 1962 with the erection of the lower level.
PERF. DURING SERV.: a. During the unstiffened state (1931-1962), the bridge experienced perceptible but not uncomfortable vibrations. b. Bottom chord lateral bracing was added to improve the performance in the stiffened state because of lessons learned from the response of the Golden Gate Bridge (S-1937A).

TABLE 2. OUTLINE OF PERFORMANCES OF FULL-SCALE BRIDGES (Cont.)

ID	NAME	DURING CONSTRUCTION				DURING SERVICE			
		VER. VIB.	TOR. VIB.	CPL. VIB.	DAMAGED	VER. VIB.	TOR. VIB.	CPL. VIB.	DAMAGED
S-1937A	GOLDEN GATE	?	?	?	No	Yes	Yes	Yes	No
	STIFFENING ELEMENT(S):	ORIG. - 25 ft. deep stiffening trusses. ADD. - Bottom chord lateral bracing was added in 1954.							
	PERF. DURING CONST.: While vibrations were observed, details are not available.								
	PERF. DURING SERV.: a. The bridge experienced large amplitude vibrations during storms on Feb. 9, 1938, Feb. 11, 1941, and Dec. 1, 1951. b. The performance improved after addition of the lateral bracing. c. During the storm of Feb. 9, 1938, the suspenders, spaced 15 in. apart, vibrated enough to slap each other.								
S-1937B	FYKESUND	?	?	?	No	Yes	No	No	No
	STIFFENING ELEMENT(S):	ORIG. - 17.7 in. deep rolled steel I-beam. ADD.- Stays were added in 1947.							
	PERF. DURING SERV.: The vertical vibrations stopped after addition of stays.								
S-1938A	THOUSAND ISLANDS, USA	Yes	Yes	?	No	Yes	No	No	No
	STIFFENING ELEMENT(S):	ORIG. - 6 ft. deep plate girders. ADD. - Temporary : Mid-span stays and main span corner stays, Permanent : Similar.							
	PERF. DURING CONST.: Temporary stays were partially effective in controlling vibrations.								
	PERF. DURING SERV.: Vibrations stopped after installation of stronger permanent stays.								
S-1938B	THOUSAND ISLANDS, Canada	Similar to S-1938A							
S-1938C	LIONS GATE	?	?	?	No	Yes	?	?	No
	STIFFENING ELEMENT(S):	15 ft. deep steel stiffening trusses.							
	PERF. DURING SERV.: The bridge vibrated vertically, with an amp. of 3 in.(single or double- ?) , under 50-60 mph winds.								
S-1939A	BRONX-WHITESTONE	Yes	No	No	No	Yes	?	No	No
	STIFFENING ELEMENT(S):	ORIG. - 11 ft. deep plate girders. ADD. - mid-span stays (1939), tower brakes (1939), tower-stays (1940), 14 ft. deep steel trusses added on top of plate girders (1946), and tuned mass dampers (1984).							
	PERF. DURING CONST.:The bridge vibrated vertically and also swung longitudinally under moderate quartering winds. Mid-span stays and tower brakes reduced frequency of occurrence of these motions but did not eliminate them.								
	PERF. DURING SERV.: a. The bridge continued to vibrate even after addition of tower stays and stiffening trusses. b. During a storm in Nov. 1968, under 70 mph winds, the bridge vibrated vertically with a max. double amplitude of 10 in. which made the car drivers uncomfortable. c. Two tuned mass dampers - located at mid-span, were added, after 45 years of service, to prevent any damage due to torsional vibrations.								
S-1939B	DEER ISLE	Yes	?	?	No	Yes	Yes	?	Yes
	STIFFENING ELEMENT(S):	ORIG. - 6.5 ft. deep plate girders. ADD. : Temporary - Mid-span stays, and main-span & side-span corner stays. Permanent - Similar to temporary stays. Later tower-stays, crisscross cable ties, and transverse bracing (in the plane of the x-section) were added.							
	PERF. DURING CONST.: Temporary stays were not fully effective in controlling vibrations.								
	PERF. DURING SERV.: After permanent stays were installed, no vibrations were reported for 15 months. Later, the vibrations resumed. During a storm (date-?), some of the stays broke. During this storm or another storm (date-?), the flanges of the side-span girders buckled. Since 1981, the performance of this bridge is being monitored by the Federal Highway Administration(FHWA).								
S-1940	TACOMA NARROWS I	Yes	?	?	No	Yes	Yes	Yes	Yes
	STIFFENING ELEMENT(S):	ORIG. - 8 ft deep plate girders. ADD.: Mid-span stays (6-1-1940) and hydraulic buffers at the towers (6-28-1940) were added before the bridge was opened (7-1-1940). Anchor stays were added in side-spans on 10-4 and 10-7-1940.							
	PERF. DURING CONST.: During erection of the deck framing, according to observers, the bridge vib. with amplitudes comparable to those for the completed structure. However, no records are available. Buffers and stays were added to improve performance.								
	PERF. DURING SERV.: a. The bridge continued to vib. in various vertical modes. b. Addition of anchor stays reduced vib.in side-spans only. However, these stays broke a week or two after installation. c. On Nov. 7, 1940, as the wind increased to 42 mph by 9.00 am, the bridge vib. vertically. At 9-30 am, a slack was observed in the mid-span stays. Around 10.00 am, the bridge started to experience tor. vibrations. Around 10.30 am, a section of the concrete roadway slab, near mid-span, fell down. Around 11.00 am, a 600 ft. portion of the deck fell down. Around 11.10 am, most of the remaining deck fell down causing the side-spans to sag 30 ft. and the top of the 434 ft. high towers to deflect shorewards a maximum of nearly 25 ft.								
S-1943A	PEACE RIVER	?	?	?	No	Yes	Yes	Yes	No
	STIFFENING ELEMENT(S):	13 ft. deep steel stiffening trusses. Lateral bracing in bottom chord - ?							
	PERF. DURING SERV.: a. Torsional vibrations, caused by winds above 40 mph, made pedestrians feel uncomfortable. b. Individual strands, arranged in an open rectangular pattern, vibrated between suspender clamps breaking a few wires. The vibrations stopped after wooden spacer blocks were installed midway between the clamps.								
S-1943B	LIARD RIVER	?	?	?	No	?	Yes	?	No
	STIFFENING ELEMENT(S):	8 ft. deep steel truss. Lateral bracing in bottom chord - ?							
	PERF. DURING SERV.: a. During Aug. 1947, the bridge vibrated with a max. double amplitude of 2 in. under steady winds of 20-25 mph. b. See S-1943A.								

TABLE 2. OUTLINE OF PERFORMANCES OF FULL-SCALE BRIDGES (Cont.)

ID	NAME	DURING CONSTRUCTION				DURING SERVICE			
		VER. VIB.	TOR. VIB.	CPL. VIB.	DAMAGED	VER. VIB.	TOR. VIB.	CPL. VIB.	DAMAGED
S-1947	BEAUHARNOIS DAM	?	?	?	No	Yes	No	No	No
	STIFFENING ELEMENT(S):	ORIG. - 7.5 ft. deep plate girder + rigid mid-span diagonal struts connecting cable bands to top of plate girders + welded floor-grid in top plane and lateral bracing in bottom plane of floor beams. ADD. - Temporary stays installed in 1950.							
	PERF. DURING SERV.:	The bridge vib. in a vertical mode, under quartering winds only, with a max. double amplitude of 13 in.							
S-1964	FORTH ROAD	?	?	?	No	?	?	?	No
	STIFFENING ELEMENT(S) :	27.5 ft. deep steel stiffening trusses with top & bottom chord lateral bracing.							
	PERF. DURING CONST.:	On Feb. 23, 1961, the full-height, free-standing north tower vibrated with a double tip amplitude of 7.5 ft. halting operations of cranes and hoistways. Vibrations stopped after installation of a temporary damping device.							
S-1966	SEVERN	?	?	?	No	?	?	?	No
	STIFFENING ELEMENT(S) :	10 ft. deep streamlined steel box girder with inclined suspenders.							
	PERF. DURING CONST.:	Vibrations were observed in the free-standing towers and later in the deck panels during erection.							

TABLE 3. AGE/LIFE-SPAN OF 26 SUSPENSION BRIDGES AS OF 1993

ID	NAME	DATE OPENED	PRESENT STATUS	AGE/LIFE-SPAN (Years)
S-1817	DRYBURGH ABBEY I	08-??-1817	Destroyed by a gale on 1-15-1818	5 months
S-1820	UNION	07-26-1820	In operation	173
S-1823	BRIGHTON CHAIN PIER	11-??-1823	Status of the bridge after the third span was damaged on 11-30-1836, and later repaired, is not known	At least 13
S-1826	MENAI STRAITS	1-30-1826	In operation	167
S-1829	MONTROSE I	??-??-1829	Demolished in 1931 because of insufficient load capacity. Replaced by a concrete cantilever bridge	102
S-1830	NASSAU	06-??-1830	Demolished in 1925 (est.). Replaced by another suspension bridge that was completed in 1926	95
S-1840	ROCHE-BERNARD	??-??-1840	Status of the bridge after it was damaged on 10-26-1852, and later repaired, is not known	At least 12
S-1849	WHEELING I	12-??-1849	Destroyed by a hurricane on 05-17-1854	4.5
S-1850	LEWISTON-QUEENSTON I	03-19-1850	Destroyed by a storm on 2-1-1864	14
S-1864	CLIFTON	12-08-1864	In operation	129
S-1869	NIAGARA-CLIFTON I	??-??-1869	Destroyed by a hurricane on 01-09-1889	20
S-1889	NIAGARA-CLIFTON II	??-??-1889	Demolished in 1897 because of insufficient load capacity. Replaced by a steel arch bridge.	8
S-1931	GEORGE WASHINGTON	·10-31-1931	In operation	62
S-1937A	GOLDEN GATE	05-27-1937	In operation	56
S-1937B	FYKESUND	??-??-1937	In operation	56
S-1938A	THOUSAND ISLANDS, USA	09-??-1938	In operation	55
S-1938B	THOUSAND ISLANDS, Canada	09-??-1938	In operation	55
S-1938C	LIONS GATE	11-??-1938	In operation	55
S-1939A	BRONX-WHITESTONE	04-29-1939	In operation	54
S-1939B	DEER ISLE	06-19-1939	In operation	54
S-1940	TACOMA NARROWS I	07-01-1940	Destroyed by a 42 mph wind on 11-07-1940	4 months
S-1943A	PEACE RIVER	07-??-1943	In operation	50
S-1943B	LIARD RIVER	07-??-1943	In operation	50
S-1947	BEAUHARNOIS DAM	??-??-1947	In operation	46
S-1964	FORTH ROAD	09-04-1964	In operation	29
S-1966	SEVERN	09-??-1966	In operation	27

APPENDIX I: REFERENCES

A. *BOOKS ON BRIDGES:*

Beckett, D. (1969). *Bridges*. The Hamyln Publishing Group, London.

Billings, H. (1956). *Bridges*. The Viking Press, New York.

Billington, D. P. (1983). *The Tower and the Bridge - The New Art of Structural Engineering*. Princeton University Press, New Jersey.

Burr, W.H. & Falk, M.S. (1912). *The Design and Construction of Metallic Bridges*. John Wiley & Sons, New York.

Casson, Sir H. (1965). *Bridges*. Taplinger Publishing Co., New York.

Condit, C.W. (1960). *American Building Art - The Nineteenth Century*. Oxford University Press, New York.

Condit, C.W. (1961). *American Building Art - The Twentieth Century*. Oxford University Press, New York.

Edwards, L. N. (1959). *A Record of History and Evolution of Early American Bridges*. University of Maine Press, Orono, Maine.

Gies, J. (1966). *Bridges and Men*. The Universal Library, Grosset & Dunlap, New York.

Hopkins, H. J. (1970). *A Span of Bridges - An Illustrated History*. Praeger Publishers, NY.

Jackson, D.C. (1988). *Great American Bridges and Dams*. Great American Place Series, The Preservation Press, Washington D.C.

Jacobs, D. & Neville, A.E. (1968). *Bridges, Canals & Tunnels - The Engineering Conquest of America*. American Heritage Publishing Co., New York.

Mock, E.B. (1949). *The Architecture of Bridges*. The Museum of Modern Art, New York.

O'Connor, C. (1971). *Design of Bridge Superstructures*. Wiley-Interscience, New York.

Outerbridge, G. & Outerbridge, D. (1989). *Bridges*. Harry N. Abrams, Publishers, NY.

Peters, T.F. (1981). *The Development of Long-Span Bridge Building*. ETH, Zurich.

Plowden, D. (1974). *Bridges - The Spans of North America*. The Viking Press, New York.

Reier, S. (1977). *The Bridges of New York*. Quadrant Press, New York.

Smith, H.S. (1953). *The World's Great Bridges*. Harper & Brothers, Publishers, New York.

Steinman, D.B. & Watson, S.R. (1957). *Bridges and Their Builders*. Dover Pub., NY.

Stephens, J. H. (1976). *Towers, Bridges, and Other Structures*. Guinness Family of Books, Sterling Publishing Co., New York.

Tyrrell, H. G. (1911). *Bridge Engineering*. Published by the author, Chicago.

Weale, J. (1855). *The Theory, Practice, and Architecture of Bridges of Stone, Iron, Timber and Wire*. John Weale, London.

Whitney, C.S. (1983). *Bridges - Their Art, Science & Evolution*. Greenwich House, Crown Publishers, New York.

B. *BOOKS ON SUSPENSION BRIDGES:*

Bleich, F., McCullough, C.B., Rosencrans, R., and Vincent, G.S. (1950). *The Mathematical Theory of Vibration in Suspension Bridges*. Bureau of Public Roads, U.S. Dept. of Commerce, Government Printing Office, Washington, D.C.

Drewry, C.S. (1832). *A Memoir on Suspension Bridges*. Longman, Rees, Orme, Brown, Green, and Longman, London.

Jakkula, A.A. (1941). *A History of Suspension Bridges in Bibliographical Form*. Bulletin of the Agricultural and Mechanical College of Texas, 4th Series, Vol. 12, No. 7, July 1, 1941.

Pope, T. (1823). *Memoir on Suspension Bridges*.

Pugsley, A. (1968). *The Theory of Suspension Bridges*. Edward Arnold, Publishers, London.

Selberg, A. (1946). *Design of Suspension Bridges*. Trondheim.

Steinman, D.B. (1949). *A Practical Treatise on Suspension Bridges - Their Design, Construction and Erection*. John Wiley & Sons, New York.

C. *BOOKS ON WIND EFFECTS:*
Houghton, E.L. and Carruthers, N.B. (1976). *Wind Forces on Buildings and Structures: An Introduction.* John Wiley & Sons, New York.
Sachs, P. (1978). *Wind Forces in Engineering.* Pergamon Press, Oxford, England.
Simiu, E. and Scanlan, R.H. (1986). *Wind Effects on Structures*, Wiley-Interscience, NY.

D. *MISC. ENGINEERING BOOKS:*
Petroski, H. (1982). *To Engineer is Human.* St. Martin's Press, New York.
Ross, S. S. (1984). *Construction Disasters: Design Failures, Causes, and Prevention.* An Engineering News-Record Book. McGraw-Hill Book Co., New York.
Schodek, D.L. (1987). *Landmarks in American Civil Engineering.* MIT Press, Cambridge, Massachusetts.
Timoshenko, S.P. and Young, D.H. (1965). *Theory of Structures* (Chapter 11 - Suspension Bridges), McGraw Hill Book Co., New York.

E. *CONFERENCE & SEMINAR PROCEEDINGS, MANUALS, AND REPORTS:*
International Symposium on Suspension Bridges. (1966). Lisbon.
Long-Span Bridges: O.H. Ammann Centennial Conference. (1980). The New York Academy of Sciences, New York.
Long-Span Suspension Bridges: History & Performance. (1979). ASCE National Convention, Boston, April 2-6, 1979. (Preprint 3590).
Samelson, H. and Fink, H. (1979). *Computer Analyses of Suspension Bridges Using Symmetry & Computer Generated Input.* Control Data Corporation.
Scanlan, R.H. (1993). "Key Mechanisms Affecting the Stability of Long-Span Bridges in Strong Wind." ASCE Metropolitan Section, Spring Seminar 1993, Structures Group, NY.
Speciality Conference on Metal Bridges. (1974). Nov. 12-13, 1974. ASCE, New York.
West, H.H. and Robinson, A.R. (1967). *A Reexamination of the Theory of Suspension Bridges.* University of Illinois, Structural Research Series, No. 322, June, 1967.
Wind Effects on Buildings and Other Flexible Structures. (1955). Notes on Applied Science, No. 11, National Physical Laboratory, Her Majesty's Stationery Office, London.
Wind Effects on Buildings and Structures. (1965). Vol.'s I &II - Proceedings of the First International Conference held at the National Physical Laboratory, Teddington, England, June 26-28, 1963, Her Majesty's Stationery Office, London.
Wind Effects on Buildings and Structures. (1968). Vol.'s I &II - Proceedings of the International Research Seminar (Second International Conference) held in Ottawa, September 11-15, 1967, Canada. University of Toronto Press.
Wind Effects on Buildings and Structures. (1971). Proceedings of the Third International Conference, Tokyo, Saikon Co., Tokyo.
Wind Effects on Structures. (1976). Proceedings of the Second USA-Japan Research Seminar on Wind Effects on Structures held in Kyoto, Japan, in September, 1974. The University Press of Hawaii, Honolulu.
Wind Loading and Wind-Induced Structural Response. (1987). An ASCE State-of-the-Art Report. Prepared by the ASCE Committee on Wind Effects, New York.
Wind Tunnel Model Studies of Buildings and Structures. (1987). ASCE Manuals and Reports on Engineering Practice, No. 67, New York.

F. *ENGINEERING PERIODICALS, JOURNALS, & TRANSACTIONS*
ASCE - CIVIL ENGINEERING MAGAZINE:
Boyer, W. H. and Jelly, I. A. (1937). "An Early American Suspension Span." Vol. 7, No. 5, May, 1937, 338-340.
Jakkula, A.A. (1933). "Various Stress Formulas Compared." Vol. 3, No. 9, September, 1933, 491-492.
Steinman, D.B. (1933). "A Generalized Deflection Theory." Vol. 3, No. 9, September, 1933, 490-491.

ASCE - JOURNAL OF STRUCTURAL ENGINEERING:

Beliveau, J.-G., Vaicaitis, R., and Shinozuka, M. (1977) "Motion of Suspension Bridge Subject to Wind Loads." Vol. 103, No. ST6, June, 1977, 1189-1205.

Billington, D.P. (1977). "History and Esthetics in Suspension Bridges." Vol. 103, No. ST8, August, 1977, 1655-1672.

Buonopane, S.G. and Billington, D.P. (1993). "Theory and History of Suspension Bridge Design from 1823 to 1940." Vol. 119, No. 3, March, 1993, 954-977.

Chang, F.-K. and Cohen, E. (1981). "Long-Span Bridges: State-of-the-Art." Vol. 107, No. ST7, July, 1981, 1145-1160.

Davenport, A.G. (1962). "Buffeting of a Suspension Bridge by Storm Winds." Vol. 88, No. ST3, June 1962, 233-268.

ASCE - TRANSACTIONS:

"Aerodynamic Stability of Suspension Bridges". (1952). 1952 Report of the Advisory Board on the Investigation of Suspension Bridges, Paper No. 2761, 721-781.

Bender, C.S. (1872) "Historical Sketch of the Successive Improvements in Suspension Bridges to the Present Time." Vol. 1, 1872.

Bleich, F. and Teller, L.W. (1949) "Dynamic Instability of Truss-Stiffened Suspension Bridges under Wind Action." Vol. 114, 1177-1222.

Bleich, F. (1952). "Structural Damping in Suspension Bridges." Vol. 117, 165-203.

Hardesty, S. and Wessman, H.E. (1939). "The Preliminary Design of Suspension Bridges." Vol. 104, 579-634.

ASCE - MISC. JOURNALS:

Dwyer, J.D. (1975)."Fifty-Year Development: Construction of Steel Suspension Bridges." Journal of the Construction Div., Vol. 101, No. CO1, March, 1975, 105-125.

Petroski, H. (1993). "Failure as Source of Engineering Judgment: Case of John Roebling." Journal of Performance of Constructed Facilities, Vol. 7, No. 1, February, 1993, 46-58.

ENGINEERING NEWS-RECORD:

"Bridge-Deck Twists Go Straight." (1983). July 14, 1983, 26-27.

Finch, J.K. (1941). "Wind Failures of Suspension Bridges or Evolution and Decay of the Stiffening Truss." March 13, 1941, 74-79.

Moisseiff, L.S. (1939). "Growth in Suspension Bridge Knowledge." August 17, 1939, 46-49.

MISC. PERIODICALS AND JOURNALS:

Berreby, D. (1992). "The Great Bridge Controversy." Discover, February, 1992, 26-33.

Petroski, H. (1993). "Predicting Disaster." American Scientist, Vol. 81, March - April, 110-113.

Stevenson, R. (1821). "Description of Bridges of Suspension." Edinburgh Philosophical Journal, Vol. 5.

Timoshenko, S.P. (1943). "Theory of Suspension Bridges." Journal of Franklin Institute, Vol. 235.

G. *ON BRIDGES INCLUDED IN THIS PAPER.*

(Note: See items "A" thru "F" for a complete description of the reference where only the author's name is listed.)

S-1817 DRYBURGH ABBEY I BRIDGE:

Bleich, F., et al. (1950), 2.

Drewry, C.S. (1832), 25-30.

Hopkins, H. J. (1970), 180-181.

Jakkula, A.A. (1941), 55-56.

Stevenson, R. (1821), 243-247.

S-1820 UNION BRIDGE:
Bleich, F., et al. (1950), 2.
Brown, Sir S. (1822). "Description of the Trinity Pier of Suspension at Newhaven."
    Edinburgh Philosophical Journal, Vol. 6, 1822, 22.
Drewry, C.S. (1832), 37-41.
Hopkins, H. J. (1970), 181-183.
Jakkula, A.A. (1941), 57.
Pugsley, A. (1968), 2.
Stevenson, R. (1821), 247-252.
Tyrrell, H.G. (1911), 209.

S-1823 BRIGHTON CHAIN PIER BRIDGE:
Bleich, F., et al. (1950), 2-4.
Drewry, C.S. (1832), 69-74.
Hopkins, H. J. (1970), 183.
Jakkula, A.A. (1941), 58.
Reid, Lt. Col. W. (1836). "A Short Account of the Failure of a Part of the Brighton
    Chain Pier, in the Gale of the 30th of November, 1836." Professional Papers of the
    Corps of Royal Engineers, Vol. 1, 99-101.
Russell, J. S. (1839). "On the Vibration of Suspension Bridges and other Structures, and
    the Means of Preventing Injury from this Cause." Transactions of the Royal Scottish
    Society of Arts, Vol. 1, January 16, 1839, 305-307.
Weale, J. (1855), Vol. 2, 150-153.

S-1826 MENAI STRAITS BRIDGE:
Baker, J.F. and Armitage J. (1937). "The Menai Suspension Bridge and Other Bridges
    Designed and Constructed by Thomas Telford in His Later Years." ASCE Civil
    Engineering, Vol. 7, No. 4, April, 1937, 277-281.
Bleich, F., et al. (1950), 5-6.
Drewry, C.S. (1832), 46-66.
Gies, J. (1966), 89-98.
Hopkins, H. J. (1970), 185-189.
Jakkula, A.A. (1941), 61-64.
Maude, T.J. (1841). "An Account of the Repairs and Alterations Made in the Structure
    of the Menai Bridge in Consequence of the Damage it Received During the Gale of
    January 7, 1839." Proceedings of the Institution of Civil Engineers, Vol. 1, Session
    January 12, 1841, 58-59.
Maude, T. J. (1841). "An Account of the Repairs and Alterations Made in the Menai
    Bridge Due to Injuries Received in Storm." Civil Engineers and Architect's Journal,
    Vol. 4, No. 44, May, 1841, 167.
"Menai Straits Suspension Bridge May Be Removed." (1927).
    Engineering News-Record, September 29, 1927.
"New Cables, Wider Broadway for Famous Menai Suspension Bridge in Wales."
    (1939). Engineering News-Record, August 3, 1939, 37.
Provis, W.A. (1841). "Observations on the Effect of Wind on the Suspension Bridge
    over the Menai Straits." Civil Engineers and Architect's Journal,
    Vol. 4, No. 45, June, 1841, 204-205.
Steinman, D.B. and Watson, S. R. (1957), 138-143.
"Telford's Menai Suspension Bridge to be Rebuilt." (1937).
    Engineering News-Record, March 18, 1937.
"The Menai Bridge." (1835). Mechanics Magazine:
    (New York), Vol. 6, No. 1, July, 1835, 12; (London), Vol. 29, No. 784,
    August, 1838, 336; and (London), Vol. 32, No. 852, December 7, 1839, 176.

S-1829 MONTROSE I BRIDGE:
Bleich, F., et al. (1950), 5.
Jakkula, A.A. (1941), 70-71.
"Montrose Suspension Bridge." (1838). The Civil Engineer & Architect's Journal,
   Vol. 1, No. 14, November, 1838, 381.
"Montrose Bridge." (1838). The Civil Engineer & Architect's Journal,
   Vol. 1, No. 15, December, 1838, 417.
Pasley C. W. (1839). "On the State of the Suspension Bridge at Montrose after the
   Hurricane of Oct. 11, 1838." Proceedings of Institution of Civil Engineers,
   Vol. 1, Session January 29, 1839, 32-33.
Rendel J. M. (1841, 1842). "Memoir on the Montrose Suspension Bridge". The Civil
   Engineer & Architect's Journal, Vol. 4, No. 50, October, 1841, 355-356; Proceedings
   of the Institution of Civil Engineers, Vol. 1, Session April 27, 1841, 122-129; and
   Journal of the Franklin Institute, Vol. 3, 1842, 116-120.
Rendel J. M. (1842). "Memoir on the Montrose Suspension Bridge." American Railroad
   Journal, Vol. 8 (New Series), No. 7, April 1, 1842, 207-212.

S-1830 NASSAU BRIDGE:
Bleich,F., et al. (1950), 6.
Jakkula, A.A. (1941), 72.
Mueller, K. (1931). "Die alte Kettenbruecke ueber die Lahn bei Nassau, Die Geschicte
   eines Brueckenbaues." Bauingenieur, Vol. 12, No. 18-19, May 1, 1931, 315-322.

S-1840 ROCHE-BERNARD BRIDGE:
Bennet, J. (1861). "Repairs and Renewal of the Roche-Bernard Suspension Bridge."
   Journal of the Franklin Institute, Third Series, Vol. 41: No 2, February, 1861, 95-108;
   No. 3, March, 1861, 145-156; No. 4, April, 1861, 227-237.
Bleich, F., et al. (1950), 6.
Hopkins, H. J. (1970), 211.
Jakkula, A.A. (1941), 109-111.
Tyrrell, H.G. (1911), 217-218.

S-1849 WHEELING I BRIDGE:
Bleich, F., et al. (1950), 6.
Dicker, D. (1969). "Aeroelastic Stability of Unstiffened Suspension Bridges."
   ASCE National Convention, Louisville, Kentucky, April, 1969.
Dicker, D. (1969). "Stiffening Systems vs. Aerodynamics."
   ASCE Civil Engineering, June, 1969, 61.
Dicker, D. (1969). "Readers Write: Suspension Bridge Stiffening."
   ASCE Civil Engineering, November, 1969, 50, 63, 69, and 72.
Gies, J. (1966), 183-185.
Hopkins, H. J. (1970), 217, 218, 227, 232.
Jakkula, A.A. (1941), 143-145.
Kemp, E.L. (1973). "Ellet's Contribution to the Development of Suspension Bridges."
   Engineering Issues - Journal of Professional Activities,
   Vol. 99, No. PP3, July, 1973, 331-351.
"Railroad Bridges-Suspension Bridges." (1856). American Railroad Journal,
   Vol. 29, No. 1044, April 19, 1856, 241-243.
Schodek, D.L. (1987), 90-92.
"Wheeling and Bridgeport Bridges - The Suspension Bridge." (1871).
   Submitted by U.S. Engineers. Report of the Chief of Engineers, Vol. 2, 1871, 405-407.

S-1850 LEWISTON-QUEENSTON I BRIDGE:
Bleich, F., et al. (1950), 6-7.
Gies, J. (1966), 18█.
Jakkula, A.A. (1941), 146-148.
"The Old and New Suspension Bridges Over the Niagara River at Lewiston, NY."
 (1899). Engineering News, Vol. 41, No. 2, January 12, 1899, 18-20.
"The Lewiston and Queenston Suspension Bridge." (1899). Engineering Record,
 Vol. 40, No. 13, August 26, 1899, 291.

S-1864 CLIFTON BRIDGE:
Beckett, D. (1969), 65-74.
Drewry, C.S. (1832), 140-142.
Gies, J. (1966), 187-188.
Jakkula, A.A. (1941), 176-179.

S-1869 NIAGARA-CLIFTON I BRIDGE:
Gies, J. (1966), 187-188.
Jakkula, A.A. (1941),184-186.
Keefer, S. (1870). "The Clifton Bridge, Niagara."
 Van Nostrand's Eclectic Engineering Magazine, Vol. 2, 318-19.
S-1889 NIAGARA-CLIFTON II BRIDGE:
Gies, J. (1966), 187.
Jakkula, A.A. (1941), 184-186.
"The Niagara Arch Bridge Swung." (1898).
 Engineering Record, Vol. 37, No. 21, April 23, 1898, 454.

S-1931 GEORGE WASHINGTON BRIDGE:
Ammann, O.H. (1953). "Present Status of Design of Suspension Bridges with Respect to
 Dynamic Wind Action." Journal of the Boston Society of Civil Engineers,
 Vol. 40, No. 3, July, 1953, 236.
ASCE Transactions. (1952), 753.
Dicker, D. (1969). "Aeroelastic Stability of Unstiffened Suspension Bridges."
 · ASCE National Convention, Louisville, Kentucky, April, 1969.
Dicker, D. (1969). "Stiffening Systems vs. Aerodynamics."
 ASCE Civil Engineering, June, 1969, 61.
Dicker, D. (1969). "Readers Write: Suspension Bridge Stiffening."
 ASCE Civil Engineering, November, 1969, 50, 63, 69, and 72.
"George Washington Bridge Across the Hudson River at New York, NY." (1933).
 ASCE Transactions. Vol. 97. Published by ASCE in collaboration with The Port of New
 York Authority.
Gould, I.P. (1958) "Design Features of Lower Deck of George Washington Bridge."
 ASCE Journal of the Struct. Div., Vol. 84, No. ST3, May, 1958, 1632-1 to 1632-22.
Jakkula, A.A. (1941), 313-324.

S-1937A GOLDEN GATE BRIDGE:
ASCE Transactions. (1952), 753, 754, 778.
Cassady, S. (1986). *Spanning the Gate: The Golden Gate Bridge.*
 Squarebooks, Mill Valley, California.
"Golden Gate Bridge is Stiffened against Wind Vibration." (1954).
 Engineering News-Record, June 24, 1954, 46-47.
Jakkula, A.A. (1941), 347-355.
"Recommended Golden Gate Bridge Bracing." (1953).
 Engineering News-Record, January 22, 1953, 22.

The Golden Gate Bridge. (1938). Report of the Chief Engineer to the Board of Directors of the Golden Gate Bridge and Highway District, California.
Vincent, G.S. (1958). "Golden Gate Bridge Vibration Studies."
ASCE Journal of the Structural Division,
Vol. 84, No. ST6, October, 1958, 1817-1 to 1817-39.
Vincent, G.S. (1959). "Correlation of Predicted and Observed Suspension Bridge Behavior." ASCE Journal of the Structural Division,
Vol. 85, No. ST2, February, 1959, 143-159.
Vincent, G.S. (1965). "A Summary of Laboratory and Field Studies in the United States on Wind Effects on Suspension Bridges." Wind Effects on Buildings and Structures.
Vol.'s I &II, Session 5, Proceedings of the First International Conference held at the National Physical Laboratory, Teddington, England, June 26-28, 1963, Her Majesty's Stationery Office, London, 488-515.

_S-1937B_ _FYKESUND BRIDGE:_
ASCE Transactions. (1952), 753, 776.

_S-1938A_ _THOUSAND ISLANDS BRIDGE, USA:_
ASCE Transactions. (1952), 752-753.
Jakkula, A.A. (1941), 356-357.
Steinman, D. B. (1938). "The Thousand Islands International Bridge."
ASCE Civil Engineering, Vol. 8, No. 6, June, 1938, 408-409.
"Three Major Bridges Completed This Month." (1938).
Engineering News-Record, August 25, 1938, 231-234.
"Two Recent Bridges Stabilized by Cable Stays." (1940).
Engineering News-Record, December 5, 1940, 56-58.
_S-1938B_ _THOUSAND ISLANDS BRIDGE, Canada:_
ASCE Transactions. (1952), 752-753.
Jakkula, A.A. (1941), 356-357.
Steinman, D. B. (1938). "The Thousand Islands International Bridge."
ASCE Civil Engineering, Vol. 8, No. 6, June 1938, 408-409.
"Three Major Bridges Completed This Month." (1938).
Engineering News-Record, August 25, 1938, 231-234.
"Two Recent Bridges Stabilized by Cable Stays." (1940).
Engineering News-Record, December 5, 1940, 56-58.
_S-1938C_ _LIONS GATE BRIDGE:_
ASCE Transactions. (1952), 753, 778-780.
Jakkula, A.A. (1941), 357-358.

_S-1939A_ _BRONX-WHITESTONE BRIDGE:_
"A Bridge to Remember." (1982).
Engineering News-Record, Editorial, January 21, 1982, 190.
Ammann, O.H. (1939). "Planning and Design of Bronx-Whitestone Bridge."
ASCE Civil Engineering, Vol. 9, No. 4, April, 1939, 217-220.
Ammann, O.H. (1946). "Additional Stiffening of Bronx-Whitestone Bridge."
ASCE Civil Engineering, Vol. 16, No. 3, March, 1946, 101-103.
ASCE Transactions. (1952), 752-754, 775-776.
"Harmonizing with the Wind." (1984). Engineering News-Record, Oct. 25, 1984, 12-13.
Jakkula, A.A. (1941), 359-360.
Reier, S. (1977), 136-138.
"Stays and Brakes Check Oscillation of Whitestone Bridge." (1940).
Engineering News-Record, December 5, 1940, 54-56.
"Weights will Stabilize Bridge." (1982). Engineering News-Record, Jan. 21, 1982, 45.

_S-1939B_ _DEER ISLE BRIDGE:_
ASCE Transactions. (1952), 752-754.
Jakkula, A.A. (1941), 360-361.
Kumarasena, T., Scanlan, R.H., and Morris, G.R. (1989).
"Deer Isle Bridge: Efficacy of Stiffening Systems." Journal of Structural Engineering,
Vol. 115, No. 9, September, 1989, 2297-2312.
Kumarasena, T., Scanlan, R.H., and Morris, G.R. (1989).
"Deer Isle Bridge: Field and Computed Vibrations." Journal of Structural Engineering,
Vol. 115, No. 9, September, 1989, 2313-2328.
Kumarasena, T., Scanlan, R.H., and Ehsan, F. (1991).
"Wind Induced Motions of Deer Isle Bridge." Journal of Structural Engineering,
Vol. 117, No. 11, November, 1991, 3356-3374.
"Two Recent Bridges Stabilized by Cable Stays." (1940).
Engineering News-Record, December 5, 1940, 56-58.

_S-1940_ _TACOMA NARROWS I BRIDGE:_
ASCE Transactions. (1952), 752-754.
Billah, K.Y. and Scanlan, R.H. (1991). "Resonance, Tacoma Narrows Bridge Failure,
and Undergraduate Physics Textbooks." American Journal of Physics,
Vol. 59, No. 2, February, 1991, 118-124.
Bowers, N.A. (1940). "Tacoma Narrows Bridge Wrecked by Wind."
Engineering News-Record, November 14, 1940, 10-12.
Bowers, N.A. (1940). "Model Tests Showed Aerodynamic Instability of Tacoma
Narrows Bridge." Engineering News-Record, November 21, 1940, 44-47.
Farquharson, F.B., Smith, F.C., and Vincent, G.S. (1949-1954). _Aerodynamic Stability_
of Suspension Bridges. (1949-1954). University of Washington Engineering Experiment
Stations Bulletin, No. 116, Parts 1-5, June 1949-June 1954.
Goller, R.R. (1965). "The Legacy of 'Galloping Gertie' 25 Years After."
ASCE Civil Engineerng, October, 1965, 50-53.
Jakkula, A.A. (1941), 362-364.
Ross, S. S. (1984), 216-239.
The Failure of The Tacoma Narrows Bridge. (1944). A Reprint of Original Reports
(Davis, Carmody, & Carew Reports), Bulletin of The Agricultural and Mechanical
College of Texas, Fourth Series, Vol. 15, No. 1, January 1, 1944.

_S-1943A_ _PEACE RIVER BRIDGE:_
Archibald, R. (1948). "Cable Vibration on Alaska Bridges."
Engineering News-Record, September 2, 1948, 79-80.
ASCE Transactions. (1952), 752-754, 780.
_S-1943B_ _LIARD RIVER BRIDGE:_
Archibald, R. (1948). "Cable Vibration on Alaska Bridges."
Engineering News-Record, September 2, 1948, 79-80.
ASCE Transactions. (1952), 752-754, 780-781.

_S-1947_ _BEAUHARNOIS DAM BRIDGE:_
ASCE Transactions. (1952), 752-755, 777-778.

_S-1964_ _FORTH ROAD BRIDGE:_
Beckett, D. (1969), 181, 182, 184, 186.
Forth Road Bridge. (1967). Institution of Civil Engineers, England.
O'Connor, C. (1971), 379-380.
Roberts, Sir G. (1965) "Design of the Forth Road Bridge." Proceedings of
Institution of Civil Engineers, Vol. 32, November, 1965, 333-405.

S-1966 SEVERN BRIDGE:
Beckett, D. (1969), 181-188.
Williams, D. T. (1967). "Severn Road Bridge." Acier-Stahl Steel, Apr. 1967, 157-162.
Fraser, R.A. and Scruton, C. (1952). *A Summarised Account of the Severn Bridge Aerodynamic Investigation.* NPL/Aero/222, National Physical Laboratory, Department of Scientific and Industrial Research, Her Majesty's Stationery Office, London.
O'Connor, C. (1971), 380-383.
Roberts, Sir G. (1969) "Severn River Suspension Bridge." (1969). ASCE Civil Engineering, August, 1969, 68-73.
Walshe, D.E. and Rayner, D.V. (1962). *A Further Aerodynamic Investigation for the Proposed River Severn Suspension Bridge.* NPL/Aero/1010. National Physical Laboratory, Department of Scientific and Industrial Research, Her Majesty's Stationery Office, London.

APPENDIX II: ABBREVIATIONS & SYMBOLS

CLEAR. - CLEARANCE
CONST. - CONSTRUCTION
CPL. VIB.- COUPLED VIBRATIONS
cps - cycles per second
ELE. - ELEMENT
est. - estimated
ft. - feet
HWY. - HIGHWAY
ID - IDENTIFICATION (MARK)
in. - inches
max. - maximum
mph - miles per hour
ORIG.- ORIGINAL
PED. - PEDESTRIAN
PERF. - PERFORMANCE
RR - RAILROAD
SERV. - SERVICE
STIFF. SYS. - STIFFENING SYSTEM
SUSP. ELE. - SUSPENSION ELEMENT
TOR. VIB. - TORSIONAL VIBRATIONS
VER. VIB. - VERTICAL VIBRATIONS
W.I. - Wrought-Iron
W.T. - WIND TUNNEL
- Item under discussion is not applicable in this case
? Information is either pending or not available

APPENDIX III : UNITS

Dimensions listed in tables are in feet units. 1 foot = 0.3048 meter.

AMBIENT AND FORCED VIBRATION TESTS
ON A CABLE-STAYED BRIDGE

Wen-Huei P. Yen[1], Thomas T. Baber [2], and Furman W. Barton[3]

Abstract

This study seeks improved prediction accuracy for dynamic parameters of long-span bridges obtained from experimental measurements. Ambient vibration tests of a long-span bridge were conducted to estimate natural frequencies, damping and mode shapes. These parameters were used to evaluate and modify a computational dynamic model for the bridge. Transient vibration tests were conducted to obtain the bridge response under the effects of two measured trucks traversing the bridge with near steady speed. With the input and output of the bridge known, a system identification method for experimental determination of the modal parameters was developed and applied to investigate the bridge's dynamic properties.

1. Introduction

In practice an analytical or computational prediction of dynamic behavior may not compare very well with measured results. One reason may be that the model parameters are inaccurate. In particular, the calculation and modeling of stiffness and damping matrices can cause significant problems. Parameter errors can be introduced by boundary condition assumptions, and assumed continuity or lack thereof between adjacent elements. If the model is sensitive to certain parameters then a small parameter error could cause large deviations in the predicted dynamic behavior. For more accurate identification of vibrating systems it is necessary to overcome the difficulties of predicting the damping factors as well as stiffness properties from measurement data that are contaminated by noise.

Long span bridge structures possess a complex dynamic response. The bridge deck dynamic behavior consists of vertical bending, torsion, lateral bending and transverse bending modes, and these modes typically interact with each other complicating the prediction of the dynamic response. Field tests of such bridges permit

---

[1]Highway Engineer, Virginia Division, Federal Highway Administration, Richmond, Virginia

[2] Associate Professor, Dept. of Civil Engineering and Applied Mechanics, Univ. of Virginia, Charlottesville, VA

[3] Professor and Chairman, Dept. of Civil Engineering and Applied Mechanics, Univ. of Virginia, Charlottesville, VA

calibration of analytical models, improve prediction accuracy, and enhance understanding of the important parameters controlling the response. During the last two decades, several ambient vibration tests [Abdel-Ghaffar and Housner(1978); Abdel-Ghaffar and Scanlan (1985a,b)] were conducted to obtain modal parameters, most of them based on wind or traffic induced vibration.

The objectives of the ambient vibration tests performed on the I- 295 cable-stayed bridge, were to obtain modal parameters by using experimental measurements collected on the actual structure for the cantilever stage of construction and for the completed bridge. The main goals were to evaluate the accuracy of several computational models (i.e. Finite Element Model) and provide information to modify the assumptions of boundary condition for improving the analytical prediction of dynamic response, and to compare the dynamic properties of the structure between the two stages of construction. The latter goal will assist in evaluating the susceptibility of such bridge superstructures to dynamic excitation during the construction stage.

Although ambient vibration tests can be used to estimate modal parameters, some limitations exist because the inputs are unknown. In order to obtain more completely defined structural properties, the input must be a known function. Then system identification techniques can be used to obtain improved estimates of structural parameters. Recently, a system identification approach was used to correct a finite element model of a high rising building and improve the accuracy [Hoff, 1989]. In view of the accuracy achieved in recent tests [Mau and Wang 1989], it appears that the frequency response data obtained have not yet been efficiently used in large structures. For instance, in symmetrical structures, the symmetry property may be used to separate closely spaced modes and to reduce the number of parameters to be estimated. Unfortunately, in structures as large as the I-295 bridge, available shakers may not be capable of generating sufficiently large excitations to permit useful signal to noise ratios to be generated. In an attempt to circumvent this difficulty, it was decided to attempt system identification using known vehicle loadings as a controlled input. Two weighted trucks, traveling at constant speed in tandem across the bridge, were used as a known input force to excite the bridge. A system identification method was developed to estimate the dynamic parameters, which were compared with the ambient vibration data.

2. The I-295 James River Bridge

The I-295 James River Bridge is a modern precast, segmentally erected, post-tensioned, cable stayed box girder bridge recently completed east of Richmond, Virginia. An elevation drawing of the continuous central seven spans of the bridge is shown in Figure 1. The bridge deck consists of twin box girders, 12 ft. (3.66 meters) deep. The side spans, all of which are 150 ft. (45.7 meters) long, were erected by the span by span method, using 20ft. (6.24 meter) long segments. The 630 ft. (192 meter) long river span was erected by the cantilever method, using 10 ft. (3.12 meter) long segments.

The river span and the two adjacent spans on either end are supported by a single plane of stay cables, arranged in a harp configuration, and passing over the precast prestressed pylons at either end of the river span. the twin boxes are joined by precast delta frame units, and a cast-in-place median closure slab. The delta frames transfer the forces from the cable stays to the twin box girders and stiffen the twin box girders, forcing them to act as a unit over the cable stayed portion of the bridge. The bridge deck cross-section at a delta frame is shown in Figure 2.

Computational Model of the Bridge

A computational model of the I-295 bridge was constructed using MSC/PAL to obtain a priori knowledge of the bridge natural frequencies and mode shapes. This model was used in selecting the reference stations in the ambient and forced vibration tests and assisted in the data analysis.

In order to obtain a better understanding of the deck motion and to determine the natural frequencies, and mode shapes more accurately, the bridge box girder was subsequently modelled by a finer mesh of plate elements. A refined MSC/PAL model adapted the stiffness properties from Lissenden's [1988] model, used a mesh of thin shell elements for the box girder and 3-D beam element for the remaining components including cables, pylons, piers, delta frames and box struts.

Taking advantage of the bridge's symmetry, the quarter bridge model shown in Figure 3 was used to reduce the DOF of the bridge model and permit greater refinement of the model within the region modeled. By applying symmetric or antisymmetric boundary conditions along the lines of structural symmetry, symmetric modes and antisymmetric modes were automatically separated by this computational model. With the animation capability of the finite element software, closely spaced modes were more easily visualized, providing insight that aided in subsequent identification of mode shape experimentally.

Boundary Conditions

Modeling structural systems by the finite element method generally requires some assumptions and idealized structural properties. In this bridge, the boundary conditions of the pylons and piers at their bases were considered as fixed. The coupling of the bridge ends with adjacent side spans were ignored, in view of the expansion joints at the ends of the spans modeled. These assumptions may be checked with the experimental results. Of the assumptions made, the most uncertainty is associated with the conditions at the end of the model, where the side spans may provide significant stiffening. Any errors introduced by this boundary condition should have little effect on the vertical bending and torsional modes but could significantly influence the calculated transverse and longitudinal frequencies, and mode shapes.

3. Experimental Procedures

Instruments

Terra Tech SSA-102 accelerometers were used in instrumenting the bridge. The accelerometers respond to frequencies down to 0 Hz, so are sensitive to gravity. Accelerometer damping is normalized in the 0.7 ± 0.15 range, and the accelerometers display a natural rolloff at about 80-85 Hz.

The data acquisition system used was a MEGADAC 2210C. 100 Hz 4 pole butterworth low pass filters built into the sample and hold modules were used, which together with the natural rolloff of the accelerometers led to an effective suppression of all input above 100 Hz. After recording, the playback system was used to transfer the recorded data to fast Fourier transform (FFT) analyzer for data analysis or to transfer the data into a personal computer hard disk for further use.

The FFT analyses were conducted using an Ono Sokki CF-350, a two channel spectrum analyzer with averaging, zooming, and selectable windows.

Ambient Vibration Testing Procedure

Several different forms of excitation were utilized to obtain data; no attempt was made in this phase of the study to ascertain the exact level of the excitation. At each experimental setup, a five minute record of bridge response to ambient wind excited conditions was obtained. During the cantilever stage measurements, additional records were generated using the instrumentation van to produce a variety of loadings on the bridge. In all cases, the data was treated as ambient data, i.e. there was no attempt to correlate the response to a measured load.

In the studies on the completed bridge, five minute ambient records were obtained, followed by several readings using two heavily loaded dump trucks, which were driven at 35 mph across the bridge.

Each cross section of the deck was instrumented with four accelerometers, three vertically oriented and one transversely oriented, as shown in Figure 4. Vertical accelerometers were placed over the outer webs of the twin box girders, to permit measured vertical motion of the bridge to be separated into contributions of the vertical and torsional modes. The vertical accelerometer located in the median, permitted the local transverse flexibility of the deck section to be measured. For the lateral modes of vibration of the bridge deck, it was only necessary to measure accelerations at the center of the deck. Stations 5 and 8 were taken as the reference stations for main span and side span monitoring respectively.

Forced Vibration testing Procedure

On the completed bridge, transient vibration experiments were conducted using the same setups as the ambient vibration test. The data acquisition system began recording data when two trucks, weighing 50 kips and 51 kips respectively, crossed the expansion joints at the ends of the bridge. The two trucks traversed the bridge at the nearly steady speed of 35 mile/hour traveling in tandem. The parameter estimation utilized the vertical responses, with station 5 taken as the reference station. Although measured data are available at 33 points, only the first setup (9 points) data was used for parameter estimation because this setup data was thought to contain more information about the dynamic response of the main span.

4. Parameter Estimation from Ambient Data

Experimental records were transferred to the Spectrum Analyzer via a digital-analog converter. To determine the natural frequencies and mode shapes from these time histories, time domain data was transformed to power spectral density functions using FFT analysis. All data was processed by using the linear summation averaging method to reduce measurement noise.

Frequencies and Natural Mode Shapes

The spectral content provides considerable information concerning the modal characteristics of the structure. Typically peaks in the frequency domain correspond to natural frequencies of the structure, although some care is needed to assure that peaks are not significantly shifted as a result of very large external masses. This is not a serious problem in structures of the size of the James River Bridge, since even very large vehicles have less than 1% of the bridge main span mass. The relative amplitudes of peaks at different transducers locations are proportional to the amplitudes of the mode shapes at that frequency. Thus estimates of modal frequencies and mode shapes are readily obtained.

Damping Factors

The damping factors were estimated by using the half-power band- width method. It can be shown that the damping factors associate with the mode i may be estimated using:

$$\zeta_i = \frac{B_i}{2f_i} \qquad [1]$$

where the f_i is the frequency of particular mode, and B_i is the half-power point band-width of the spectral peak of f_i. In order to obtain reasonable estimates of the damping, a number of data points must be available in the spectrum in the vicinity of the peak for a particular modes. To maximize the data between the band-width of a particular mode, averaging and zooming functions must be used in obtaining the band-width data at each particular mode, and closely spaced symmetrical and antisymmetrical modes must be separated by combining data.

Parameter Estimation from Forced Vibration

Recently, frequency domain methods to directly improve the system matrices of the analytical model have been developed by Natke (1987), Cottin et al. (1984), Hoff (1988), and Fritzen (1985). A generalized multiple inputs and multiple outputs (MIMO) modal parameter estimation algorithm for incomplete modal testing was written by Craig and Blair (1985) . The present study develops a new direct modal parameter estimation method to fit an incomplete and closely-spaced-modes model.

For an incomplete computational model, m modes are identified ($m \leq N$). Then the system model is approximated by

$$H(\omega) = \sum_{r=1}^{m} \frac{\psi_r \psi_r^T}{(-\omega^2 M_r + i\omega C_r + K_r)} \qquad [2]$$

where ψ_r is the mode shape amplitude vector associated with mode r.

If the outputs measured are the accelerations of the structure, then

$$\ddot{X}(\omega) = -\omega^2 H(\omega) F(\omega) \qquad [3]$$

Assume that the transfer function can be decomposed into the form

$$-\omega^2 H(\omega) = \psi[/T_r(\omega)/]\psi^T \qquad [4]$$

where

$$T_r(\omega) = -\omega^2(-\omega^2 M_r + i\omega C_r + K_r)^{-1} \qquad [5]$$

Equations [4] and [5] imply that the structure has proportional damping. Assume that the responses are measured at n physical DOFs. If it is also assumed that the structure can be described accurately over the frequency range of interest by m DOFs ($m \leq n$), then for the case where the measured response locations are selected as the active degrees of freedom, equation [3] can be partitioned into active and non-active equations. Thus,

$$\begin{Bmatrix} \ddot{X}_a(\omega) \\ \ddot{X}_n(\omega) \end{Bmatrix} = \begin{bmatrix} \psi_a \\ \psi_n \end{bmatrix} [/T_r(\omega)/] \begin{bmatrix} \psi_a^T & \psi_n^T \end{bmatrix} \begin{bmatrix} F_a(\omega) \\ F_n(\omega) \end{bmatrix} = \begin{bmatrix} [I] \\ \psi_n\psi_a^{-1} \end{bmatrix} \psi_a[/T_r(\omega)/]\psi_a^T F_{aa}(\omega) \qquad [6]$$

where

$$F_{aa}(\omega) = F_a(\omega) + \psi_a^{-T}\psi_n^T F_n(\omega) \qquad [7]$$

Thus the equations relating the active accelerations to the forces is

$$\left[\psi_a[/T_r(\omega)/]\psi_a^T\right]^{-1} \ddot{X}_a(\omega) = F_{aa}(\omega) \qquad [8]$$

Define a parameter matrix $P^T = (P_m^T, P_c^T, P_k^T)$ of coefficients used to modify the modal quantities. Then modified modal parameters are

$$[/M_r/] = P_m^T [/\overline{M}_r/] \quad [/C_r/] = P_c^T [/\overline{C}_r/] \quad [/K_r/] = P_k^T [/\overline{K}_r/] \qquad [9]$$

where $[\overline{M}_r]$, $[\overline{K}_r]$ are the a priori estimates of the modal mass and stiffness obtained from the computational model of the structure, and $[\overline{C}_r]$ are modal damping estimates obtained from the ambient vibration test. The initial values for the diagonal elements of the parameter matrix are all 1, with the actual values to be estimated by the identification method. Thus, the system modal model $H(\omega)$ is a function of parameter P.

Estimation Method

For the estimation of the system modal parameters matrix P a loss function must be defined so that these parameters will be optimized in the estimation. In this study, a least squares method of minimizing the residual between computational and measured model has been used. Thus, the loss function, $J(\varepsilon)$ consists of a sum of squares of the residual vector ε. When operating in the frequency domain the residuals are complex and are defined by

$$\varepsilon = \alpha_m - \alpha_c(P) \qquad [10]$$

The α_m are measured quantities of the real system to be investigated, the α_c are the corresponding quantities of the associated mathematical model, which adopt the a priori information from the computational model and depend only on certain parameters P. The loss function is given by

$$J(p) = \varepsilon^{*T}(p)\varepsilon(p) \tag{11}$$

Where * indicates complex conjugate and T represents transpose.

The input residual method was used since it leads to equations that are linear in the parameters P. Then

$$\varepsilon_i(p) = F_m - F_c(p) \tag{12}$$

where

$$F_c(p) = (\frac{-1}{\omega^2})H(p)^{-1}\ddot{X}_m \tag{13}$$

The input residuals are defined as the difference between the measured input F_m and the input F_c calculated from the computational model $H(p)$ (which depends only on parameters p) with the measured output X_m.

5. Comparison of Experimental and Computational Results

Computational Model and Ambient Vibration Testing Data

Comparisons of predicted natural frequencies obtained from the computational model and experimental results in the two different stages, i.e the cantilever stage and completed bridge, are shown in tables 1, and 2 respectively. In constructing these tables, the shapes of the experimentally determined and computed modes were compared to verify that the frequencies being tabulated were of the same modes. In general, natural frequencies calculated by the finite element models are consistent with the experimental values. In some fundamental modes, they are almost identical, but computed frequencies appear to underestimate the experimentally determined value in some of higher modes. This suggests that the stiffness may be under-estimated and may need to be revised since it is expected that the mass matrix is more accurate than the stiffness matrix because the mass is dependent upon the density and volume of the material, and was obtained more accurately that stiffness.

In comparing these tables, it should be noted that the cantilever stage measurements were taken in the main span only, thus side span modes were not recorded at this stage. The vertical bending modes and lateral modes' frequencies increased when the bridge was completed, as would be expected. The fundamental torsional frequency did not change much between the two stages, but additional symmetrical torsional modes appeared.

Comparisons of the vertical bending and torsional mode shapes between the two models for the cantilever and complete stages are shown in Figures 5 and 6 respectively. In the main span, the first vertical bending mode of the completed bridge appears to have less main span curvature than the computed mode. The remaining vertical bending modes and torsional modes are consistent with this finite element model. Transverse mode shapes participated to a much smaller degree, and were only approximately identified (Yen et al, 1992)

Comparison of Ambient and Forced Vibration Testing Results

The modal stiffness and damping obtained from the forced vibration test are compared to the results which were obtained from the ambient vibration test in Table 3. In Table 3, c_i is the ith modal damping parameter estimated, while k_j is the jth modal stiffness parameter. Columns 2 and 3 show the values of the parameters obtained from the ambient vibration study and the system identification procedure respectively. The final "Mode Description" column of the table indicates the mode for which the parameters were estimated.

The results from the least squares system identification method indicate that in the higher modes the damping values are larger than the values which were estimated from ambient vibration tests. This is consistent with the practical problem of measuring damping.

If the input disturbance are negligible then the variation of the higher modes damping indicates that the structure system response caused by different kinds of excitations may have different levels of damping. These results agree with Kawashima's (1990) study which indicated the damping of long-span bridges caused by two different levels of excitations (ambient and earthquake condition) revealed different damping levels. Another possibility is that the data used for parameter estimation are contaminated by the input process noise, thus the higher modes damping are not accurate by this identification procedure. Finally, this deviation may be caused by the proximity of the trucks' natural frequencies to the higher modal frequencies of the bridge and the fact that the trucks are highly damped. The system identification accuracy could be improved by applying an optimal input to increase the output response level so that the noise to signal ratio is decreased.

6. Conclusions

This paper describes ambient and forced vibration testing of the I-295 James River Bridge, aimed at developing a method to estimate system properties such as modal damping and stiffness for a cable-stayed long-span bridge. The dependability of a finite element model was evaluated by a field response data analysis from ambient vibration testing. Excluding the first longitudinal mode, good agreement as observed between the computational model's and field test data's natural frequencies and mode shapes. This exceptional mode may be caused by the assumption that of the boundary condition at the ends of the model are free to translate longitudinally. In reality the sides span may provide significant stiffening. Damping factors obtained from the half-power method are near 1% which is around the average damping of long-span bridges from previous tests.

Modal stiffnesses obtained using the identification procedure agreed well with the results of ambient vibration testing. The modal damping parameters in the higher modes are three to four times larger than lower modes. Further study is needed before it can be said that the damping is increased by this kind of excitation.

Acknowledgement

Support for the work was provided by Grant Number CES-8715463 from the National Science Foundation. Additional support of the work was provided through the Virginia Transportation Research Council. Assistance in the field work by W. T. McKeel Jr., James W. French, Jean-Marie Roque and Paul Duemmel is gratefully acknowledged.

References

Abdel-Ghaffar, A. M.and Housner, G. W. (1978) "Ambient Vibration Tests of Suspension Bridge" J. Eng. Mech. Div. ASCE, 104, EM5, 983-999.

Abdel-Ghaffar, A. M. and Scanlan, R. H. (1985a) "Ambient Vibration Studies of Golden Gate Bridge: I. Suspended Structure," J. Eng. Mech. Div.ASCE, 111, 463-482.

Abdel-Ghaffar, A. M. and Scanlan, R. H. (1985b) "Ambient Vibration Studies of Golden Gate Bridge: II. Pier-Tower Structure," J. Eng. Mech. Div. ASCE, 111, 483-499.

Cottin, N., Felgenhauer, H-P. and Natke, H. G. (1984), "On the Parameter Identification of elas-tomechanical Systems Using Input and Output Residuals," Ingenieur-Archiv, 54, 378-387.

Craig, R. R. Jr.and Blair M. A., (1985), "A Generalized Multiple-Input, Multiple-Output Modal Parameter Estimation Algorithm," AIAA Journal, 23, 6, 931-937.

Fritzen, C-P (1985), "Identification of Mass, Damping, and Stiffness Matrices of Mechanical Systems," Transaction of ASME June, Paper No. 85-DET-91.

Hoff, C. (1989) "The Use of Reduced Finite Element Models in System Identification," Earthquake Engineering and Structural Dynamics, 18, 875-887.

Kawashima, K., Unjoh, S. and Azuta, Y., (1990), "Analysis of Damping Characteristics of Cable Stayed Bridge Based Strong Motion Records," Proceedings, 6th U. S.-JAPAN Bridge Engineering Workshop, May 1990 sec.5.

Lissenden, C. J. III, (1988), *Dynamic Modeling of a Cable-Stayed Bridge During Construction*, M. S. Thesis, Department of Civil Engineering, The University of Virginia, Charlottesville, VA.

Mau, S.T. and Wang, S. (1989), "Arch Dam System Identification Using Vibration Test Data," Earthquake Engineering and Structural Dynamics, 18, 491-505.

Natke, H.G. (1987),"Application of system identification in Engineering," CISM Course, New York.

Yen, Y-H. P., Baber, T. T., and Barton, F. W., (1992), *Modal Identification of Bridge Structures with a Case Study on the I- 295 James River Bridge*, Technical Report No. UVA/526691/CE92/101, on NSF grant CES-8715463.

Table 1. Selected Computed and Measured Frequencies - Cantilever Stage

Mode	Quarter Bridge Model (Hz)	Field Test Data (Hz)	Damping Ratio (%)
(a) Symmetric Modes - Frequencies and Damping Ratios			
First Longitudinal Mode	0.419		
First Vertical Bending Mode	0.477	0.470	1.25
Second Vertical Bending Mode	1.503	1.6200.41	
Third Vertical Bending Mode	3.510	3.875	0.45
(b) Antisymmetric Modes - Frequencies and Damping Ratios			
First Lateral Bending Mode	0.424	0.440	0.71
First Torsion Mode	1.047	1.132	0.72
Second Lateral Bending	2.250	2.265	0.31
Second Torsion Mode	3.047	3.246	0.19

Table 2. Selected Computed and Measured Frequencies - Completed Bridge

Mode	Quarter Bridge Model (Hz)	Field Test Data (Hz)	Damping Ratio (%)
(a) Sym./Sym. Modes - Frequencies and Damping Ratios			
First Vertical Bending	0.5677	0.5664	0.71
Third Vertical Bending	1.9910	2.1125	0.52
First Transverse Bending	3.151	3.025	0.85
(b) Sym./Antisym. Modes - Frequencies and Damping Ratios			
First Longitudinal Mode	0.423		
Second Vertical Bending	1.120	a1.175	0.42
Side Span 2d Vert. Bend.	2.963	2.885	1.25
Fourth Vertical Bending	3.232	3.275	0.64
(c) Antisym./Sym. Modes - Frequencies and Damping Ratios			
1st Lateral Bending	0.558	0.550	0.93
First Torsion	1.047	1.085	0.51
Third Torsion		3.425	0.525
(d) Antisym./Antisym. Modes - Frequencies and Damping Ratios			
Second Lateral Bending	1.754	1.750	0.75
Second Torsion	2.088	2.118	0.52

Table 3. Modal Parameters Comparison - Completed Bridge

	Ambient	Sys. Id.	Mode Descript.
c1	0.71	0.865	1st V.B.
c2	0.51	0.822	1st Tors.
c3	0.42	0.722	2nd V.B.
c4	0.52	1.023	3rd V.B.
c5	0.52	0.467	2nd Tors.
c6	0.64	1.976	4th V.B.
c7	0.52	2.817	3rd Tors.
k1	0.566	0.572	1st V.
k2	1.085	1.065	1st T.
k3	1.175	1.212	2nd V.
k4	2.112	2.212	3rd V.
k5	2.210	2.228	2nd T.
k6	3.275	3.195	4th V.
k7	3.425	3.303	3rd T.
c n : damping ratio in % .			
kn : natural frequency in hertz.			

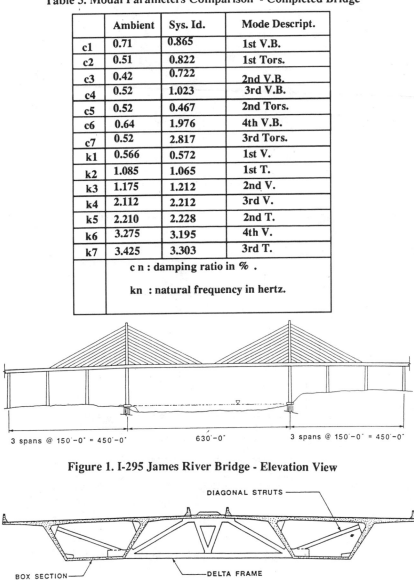

3 spans @ 150'-0" = 450'-0"   630'-0"   3 spans @ 150'-0" = 450'-0"

Figure 1. I-295 James River Bridge - Elevation View

DIAGONAL STRUTS

BOX SECTION   DELTA FRAME

Figure 2. Twin Box Girder Cross-section at Delta Frame

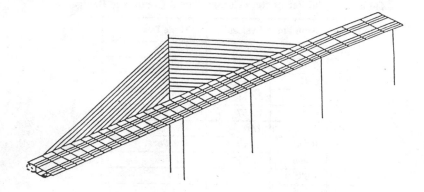

Figure 3. Quarter Bridge Finite Element Model

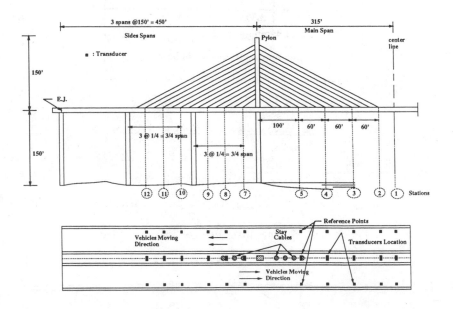

Figure 4. Location of Accelerometers on Completed Bridge

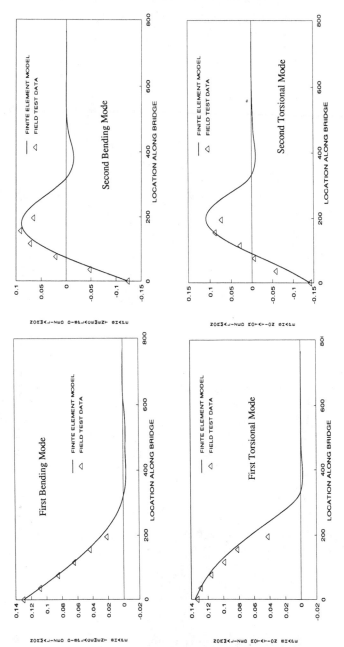

Figure 5. Computed and Measured Mode Shapes - Cantilever Stage

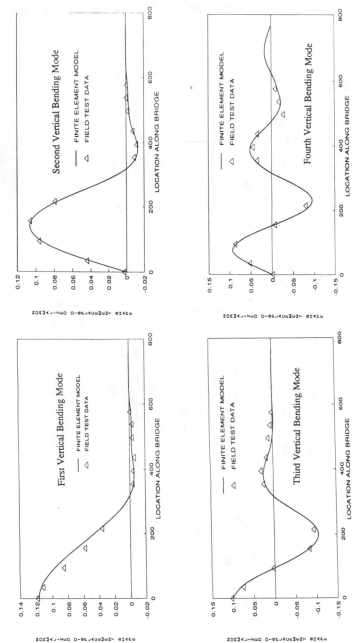

Figure 6. Computed and Measured Mode Shapes - Completed Bridge

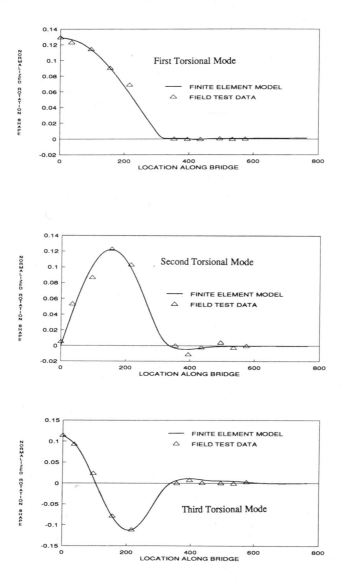

Figure 6. Computed and Measured Mode Shapes - Completed Bridge

SUBJECT INDEX
Page number refers to first page of paper

AUTHOR INDEX
Page number refers to first page of paper